14.98

# CRYSTAL
## IDENTIFIER

# CRYSTAL
## IDENTIFIER

### PETER DARLING

**MALLARD PRESS**

**DEDICATION**

**I dedicate this book to Helen, who made sense of my
turmoil and in the process made an engineer very happy.**

MALLARD PRESS

An imprint of BDD Promotional
Book Company, Inc.,
666 Fifth Avenue
New York, N.Y. 10103

Mallard Press and its accompanying design and
logo are trademarks of BDD Promotional Book
Company, Inc.

First published in the United States of America
in 1991 by the Mallard Press

ISBN 0–7924–5511–8

A QUINTET BOOK

This book was designed and produced by
Quintet Publishing Limited
6 Blundell Street
London N7 9BH
Creative Director: Terry Jeavons
Art Director: Ian Hunt
Designers: James Lawrence, Peter Radcliffe
Project Editor: Sally Harper
Editor: Ken Gleeson

Typeset in Great Britain by
Central Southern Typesetters, Eastbourne
Manufactured in Hong Kong by
Regent Publishing Services Limited
Printed in Hong Kong by
Leefung-Asco Printers Limited

# CONTENTS

*The Crystal* . . . . . . . . . . . . . . . . . . . . . . . . . . . . . . . . . . . . . . . . . . . . . . *6*

*Distinguishing Features of Crystals* . . . . . . . . . . . . . . . . . . . . . . . *10*

*Crystal Identifier* . . . . . . . . . . . . . . . . . . . . . . . . . . . . . . . . . . . . . . . . . . *20*

*Glossary* . . . . . . . . . . . . . . . . . . . . . . . . . . . . . . . . . . . . . . . . . . . . . . . . . . *77*

*Index* . . . . . . . . . . . . . . . . . . . . . . . . . . . . . . . . . . . . . . . . . . . . . . . . . . . . *79*

# The Crystal

**R**egular, beautiful, exact – crystals have excited imagination and desire for thousands of years. Created by chance in an unbelievably hostile environment, crystals have a form precise enough to delight scientists, colours bold and variable enough to inspire artists, and a chemical make-up as unpredictable and intriguing as the weather.

It is no wonder, then, that their beauty, regularity and suggestive powers have guided, aided and haunted people since the dawn of time. Here was something that could be carried, would not run away or rot – and could be easily hidden.

Without even a basic understanding of nature, early civilizations were unsurprisingly driven to imbue crystals with supernatural properties, ready to discharge their powers in the service of their master. They are also believed to have been amongst the first materials used by primitive people, who chipped and fashioned crystals before securing them to branches to form crude but effective weapons.

Simultaneously, they have been used for centuries to ward off evil spirits and cure all manner of ailments; to bring order to chaos and strength to the weak. It is for this reason that the crowns of kings and bishops, tokens of love and devotion, amulets against suffering and good luck charms have frequently been embellished with crystals, usually in the form of gemstones. On a more practical level, gemstones have always proved to be a safe and convenient way of not only storing but also displaying wealth.

Today, of course, we have a more complete and scientific understanding of crystals.

They can most simply be defined as naturally occurring inorganic minerals which take on a uniform shape within a specific structure. Their physical and chemical properties are defined by the elements which make up each individual crystal type, and by the environment in which they are formed.

Although everybody assumes that crystals are bright, shiny and regular, they actually come in a variety of colours and shapes. But perhaps the most astonishing source of their uniqueness is the fact that their usefulness, value and relative abundance bear no relationship to one another at all.

However, the greater scientific knowledge of the present century has not detracted from the interest which still persists in what are perceived to be their hidden powers. Many still hold the belief that crystals, in their elegance, purity and stability, contain the strength to guide or correct wandering or wayward spirits.

And there is little doubt, wherever one draws the line between scientific fact and mystical fancy, that they have often provided hope and relief to those in mental or physical distress. Equally, it is clear that the

*BELOW This 18th-century soapstone figure is Shoulao, the Chinese god of longevity: one of the many associations between crystals and good luck charms.*

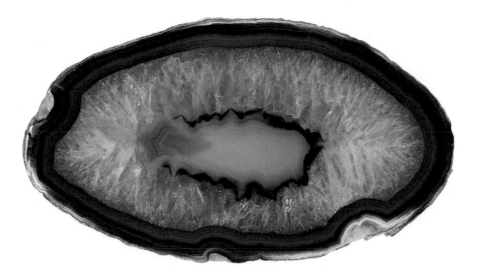

recent renewed interest in the medical benefits of crystals is due to a certain disenchantment with established medical practices.

In much the same way that homeopathic medicines are being used as a supplement or alternative to conventional treatments, crystals have maintained their place alongside the scalpel in the surgery or the bottle of pills in the medicine cabinet.

One of the most common crystals is rock salt or halite which, besides being used in food preparations, is also the major source of the valuable elements of soda, sodium and hydrochloric acid – without which the chemical industry could not exist.

In addition to this, salt deposits hundreds of feet below the surface of the waters in the Gulf of Mexico have evolved into massive dome structures whose interest goes beyond the realms of structural geology alone. These deposits – adding the tangibility of value to the fancy of spiritualism – are now one of the world's most abundant sources of salt, sulphur, petroleum and natural gas.

## CRYSTAL FORMATION

How is it that something as perfect, as beautiful and as useful as a crystal can come from the Earth where it is usually found amongst unimpressive looking rocks?

With few exceptions, minerals are not created, but grow from a tiny nuclei within a solution into crystals. It is still unclear what causes this initial nuclei to occur; it may be that atoms normally being agitated at very high frequency exist by chance for an instant in the exact orientation that they would assume if they were solid. This chance occurrence may be enough to form the nuclei and start the whole crystal growth process.

The growth of crystals at this stage is similar to that of a pearl which develops layer by layer from a grain of sand within an oyster. It is hard to believe that crystals form one atom at a time, but this process takes place in three dimensions and is repeated thousands of times a second. Depending on the size of the crystal, they can take anything from a day to a year to be completed.

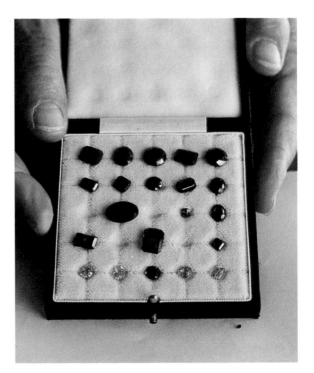

ABOVE *A selection of cut sapphires from Australian mines; sapphires are Australia's most profitable crystal.*

*OPPOSITE Like most crystals, this eruption of lava at Kilauea Iki, Hawaii, owes its existence to activity originating deep within the Earth's crust.*

## CRYSTAL SOLUTIONS

Crystals are formed from gases, molten rock or aqueous solutions which are usually created far below the Earth's surface. Sometimes they are reformed from previously solid material which has been heated and pressurized until it liquefies, only resuming a solid state when the pressure and/or the heat source is removed. Such re-formed crystals may not resemble their original form due to the addition or subtraction of elements, or because of a change in the heat and/or pressure of the growth environment.

Crystals are often formed as a result of hydrothermal action. A super-hot solution heavily charged with chemical elements is forced by high temperature through micro-cracks and veins. As this solution is displaced its temperature and pressure are dissipated. When the conditions are right and less 'turbulent', crystals will grow out of the solution.

Crystals are also formed from solutions of elemental salts which become steadily more concentrated as they exceed their saturation level. This usually happens when the inlet or outlet of an inland lake or sea is altered by geological uplift and the resulting solution becomes progressively more concentrated as natural evaporation occurs, until saturation is achieved and crystallization follows.

## GROWTH RATE

The size of crystals depends on the rate of growth: the slower they grow, the larger they will be. This may be due to evaporation, a cooling off of the solution, or the bleeding off of pressure. Sometimes the results of slow crystal growth are spectacular; in Brazil a beryl crystal weighing 200 tons (179 tonnes) was found and in Siberia a milky quartz crystal of 13 tons (11.6 tonnes) was uncovered. Slow growth rate also causes the crystal to be shaped more regularly because the atoms have more time in which to assume some orderly arrangement.

The faces of a crystal do not necessarily grow at the same speed. This difference in the rate of face growth, if very pronounced, will result in elongated crystals. A faster accumulation of atoms in one direction, due possibly to a strong electrical attraction, will be exaggerated as the atoms are being attracted to a small face-end rather than to a long side.

Similarly if a crystal grows from a solid base, as is so often the case with amethysts and halite, only half of the crystal is in a position to receive new unit-cells. It is astonishing that these crystals are not more frequently malformed and irregular in shape.

Even under ideal conditions, few minerals do not form regular shapes, but simply assume a more rigid form of the liquid phase. When such minerals are formed from a gel they are usually composed of tiny crystals, which, given time and stable conditions, will join together to form solid masses or aggregates.

*ABOVE The Oppenheimer diamond, held at the Smithsonian Institution, Washington, USA, weighs in at 254 carats.*

# Distinguishing Features of Crystals

### CONSISTENCY OF ANGLES AND SHAPES

The regularity with which crystals form was first noted in 1669 by Nicolaus Steno, the Danish scientist, who showed that at a specific temperature and pressure all crystals of the same substance possess the same angle between corresponding faces. This is known as the law of constancy of interfacial angles. The law was later clarified in the early part of the 19th century by Reinhard Bernhardi who showed that the angle referred to was taken at right angles to the faces (or at a projection of the faces) towards the central point of the crystal.

One of the truly fascinating aspects of crystals is that despite their seemingly limitless shapes they belong to one of only seven classifications of symmetry. These are categorized by their crystal axes, the angles at which the axes radiate from the crystal's central point of intersection, and the planes in which these axes lie.

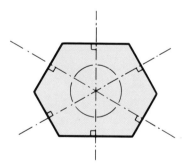
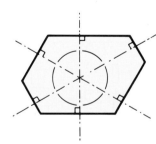
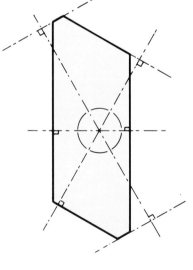

Different sections through the main axis of quartz crystals. The crystal to the left has developed regularly whilst the other two have developed unequally but still comply to the law of constancy of interfacial angles.

| CLASSIFICATION | FACE SHAPE AND AXIS ORIENTATION | FORM |
|---|---|---|
| **Isometric** | | All three axes are the same length and are at right angles to each other. |
| **Tetragonal** | | Three axes which are at right angles to each other. The two on the same plane are equal in length while the third is perpendicular to this plane. |
| **Hexagonal** | | Three of the four axes are in a single plane and radiate out equally from a central point. The fourth axis is at right angles to this plane and is unequal in length to the others. The crystal has six distinctive planes of symmetry parallel to the long axis. |
| **Trigonal** | | Similar to the hexagonal system in that there are three equal axes radiating from a single point in the same plane. A fourth axis is at right angles to this plane. There are three distinct planes of symmetry parallel to the long axis. |
| **Orthorhombic** | | Three axes of unequal length set at right angles to one another. |
| **Monoclinic** | | The prism has inclined top and bottom faces. There are three axes of unequal length: two are at right angles to each other with the third set at an incline to the plane of the other. |
| **Triclinic** | | Three axes of unequal length set at three different angles to one another. Three pairs of faces. |

Although any crystal will belong to one of these seven classifications it should be noted that crystals very seldom occur as 'copy-book' examples of their crystal model. Many minerals occur in a combination of mineral forms. These are called polymorphs of which calcite has 80. Most crystals favour one or two classic forms.

## CRYSTAL HABIT

The number and orientation of faces that a crystal has is a function of the atomic structure of the elements and the environment in which the crystal solution solidifies. The environment, which is the balance between the temperature, pressure and chemical saturation of the elements within the solution, will have a marked effect on the shape of the crystals that eventually emerge.

Although in theory there is no limit to the possible shapes that a crystal will assume – and they can be spiny, long, columnar, squat, square, tabular, plate-like, etc – generally, crystals of a particular type tend to grow in a regular fashion. This is referred to as 'crystal habit', and is used to describe the size and the shape most frequently taken by the crystal.

Crystal habit is the result of a compromise between the way the solution would naturally crystallize out, given ideal conditions, and the environment in which the solution most often occurs.

*LEFT This fine example of feldspar variety adularia from Switzerland shows clearly the striations that may occur on the surface of a crystal.*

## STRIATIONS

Striations on crystal faces are ridges, furrows or linear marks believed to be due to an oscillation of growth between two crystal orientations. Such oscillatory growth tends to lead to less well defined crystal edges. Rounded faces are a characteristic of such behaviour, and are especially common in tourmaline crystals.

## CHEMICAL IMPURITIES

The presence of minute chemical impurities – whilst possibly acting as the initial nuclei mentioned earlier – is often thought to cause a slowing up of crystal growth, which leads to a different crystal habit from that which normally occurs. Impurities contained within crystals often manifest themselves as inclusions or as distinctive colours. In fact these 'impurities' often have a beneficial effect on crystals, turning an otherwise undistinguished crystal into a unique object of beauty.

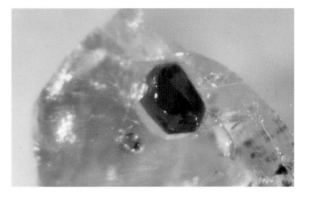

*ABOVE Inclusions can take dramatic forms, such as this garnet encapsulated by diamonds. The garnet was formed some 3,000 million years ago, while the diamond formation is relatively young – 90 million years old. This extraordinary sample is from southern Africa.*

*LEFT Schists, such as this kyanite, are rocks which contain mineral deposits in parallel or sub-parallel veins.*

*BELOW A crystal sample shown under the polarizing microscope, showing clearly the areas of different mineral types that appear as different colour zones.*

## COLOUR

Crystals with similar structures which differ in chemical composition (even in minute detail) are often quite different colours and shapes. This can lead to a marked variation in value, especially in gemstones. Classic examples of this are provided by the ruby and the sapphire, both members of the corundum group; rubies owe their colour to traces of chrome, while sapphires owe their colour to minute quantities of iron and titanium. Because of the relative scarcity of chrome compared to iron, rubies are more valuable than sapphires.

Colour is produced by reflected light. White light is made up of electromagnetic pulses with seven different wavelengths; these produce the colours of the spectrum (red, orange, yellow, green, blue, indigo and violet). These combined colours enter the crystal and some are absorbed (or partly absorbed) by an element within the crystal. It is the combination of these reflected wavelengths that we see as colour. The metal elements chrome, cobalt, copper, iron, magnesium, nickel, titanium and vanadium are common wavelength absorbers and thus colour creators.

If all wavelengths pass through the crystal, it appears to be colourless; if all wavelengths are absorbed it appears to be black; if all wave-

*ABOVE Some crystals demonstrate an optical characteristic called birefringence, in which an object viewed through the crystal is seen double. In this instance the crystal is calcite from Iceland.*

lengths are absorbed to the same degree, the crystal will appear dull or grey. Absorption and colour can also be influenced by the distance that light has to travel through a stone. This distance factor is useful to gem cutters, who make light coloured stones thicker to enhance their coloration and hollow out the back of deep coloured stones, such as red almandine garnet, for the opposite effect.

Artificial light can have a different effect to natural light: it enhances the beauty of emeralds and rubies, but diminishes the appeal of sapphires. The most obvious colour change occurs in alexandrite, which is green in natural light and red in artificial light.

## HARDNESS

Hardness, an inherent and easily determined characteristic of a mineral, is measured using a resistance to scratching method known as 'Mohs' Scale', named after Freidrich Mohs (1773 – 1839), a Viennese mineralogist. Frustrated by the existing method of describing minerals in such scientifically imprecise terms as 'soft' or 'hard' (with the prefixes of 'medium', 'very' or 'extreme'), Mohs decided that a more scientific approach to hardness measuring was required. If mineralogy was to be considered a true science it would have to meet science's exacting standards in nomenclature to begin with.

Accordingly, Mohs chose 10 minerals of varying degrees of hardness, and allocated each a 'hardness number' ranging from 1 as the softest to 10 as the hardest. Each mineral in the series is capable of scratching the mineral below it in the scale, as well as itself being scratched by the one above. Those with a hardness of 1 to 2 are called soft, 3 to 6 are termed medium-hard, 6 to 8 are hard, and 8 to 10 have 'precious-stone' hardness status. The scale that this provides is known as Mohs' scale.

| MOHS' SCALE | Comparison Mineral | Mineral Test | Rosiwal's Grinding Hardness |
| --- | --- | --- | --- |
| 1 | Talc | Powdered by finger nail | 0.03 |
| 2 | Gypsum | Scratched by finger nail | 1.25 |
| 3 | Calcite | Scratched by copper coin | 4.5 |
| 4 | Fluorite | Easily scratched by pocket knife | 5.0 |
| 5 | Apatite | Just scratched by pocket knife | 6.5 |
| 6 | Orthoclase | Scratched by steel file | 37 |
| 7 | Quartz | Scratches glass window | 120 |
| 8 | Topaz | Easily scratches quartz | 175 |
| 9 | Corundum | Easily scratches topaz | 1,000 |
| 10 | Diamond | Cannot be scratched | 140,000 |

Traditionally, a geologist would carry samples of most of the minerals in the Index in order to be able to field-test rocks. Nowadays, it is possible to buy a set of what are known as hardness pencils; these consist of splinters of the appropriate mineral set into a metal holder with the hardness value clearly indicated on the holder. Care should be taken when conducting a hardness test that only a sound surface is tested, and then only with a sharp point of the test mineral.

The advantage of Mohs' scratch hardness test is its simplicity and ease of application. However, it is an empirical scale and does not bear any relationship to hardness in the strict scientific sense of that term. This can be seen by consulting the right-hand side of the Hardness Table which lists the Rosiwal cutting resistance or grinding hardness of the materials used in Mohs' scale. The Rosiwal scale was devised by August Karl Rosiwal (1860–1923), and is a scientific scale of hardness as opposed to Mohs' empirical scale.

The hardness of a material depends on the atomic bonding of the crystal structure, since these bonds can vary depending on their crystallographic direction. Hardness variations are displayed by striated, laminated or weathered crystals, among others. The most celebrated example of hardness variation occurs in the kyanite crystal which displays a hardness of 4 to 5 along its crystal axis but a value of 6 to 7 across the crystal axis.

## LUSTRE

Lustre is concerned with a mineral's surface and the intensity with which that surface reflects light. A mirror would have perfect lustre, described as 'brilliant'; whilst most earthy materials, such as clay, are matt in appearance and are described as 'dull'.

The two main types of lustre are metallic and non-metallic. Whereas freshly cut metals and most sulphide materials have metallic lustres, there are a host of terms used to describe those found in non-metallic materials. 'Adamantine' is a brilliant lustre usually found in such crystals as diamond and cassiterite. 'Vitreous' is the most common form of lustre and can be described as glass-like, and is commonly found in quartz and beryl. A 'resinous' lustre is displayed by sulphur, whilst fibrous minerals like gypsum and asbestos display a 'silky' lustre. Minerals with a 'waxy' lustre reflect only a small portion of the light hitting their surface, while minerals with an 'earthy' or 'dull' lustre are normally either aggregates or have rough weathered surfaces (examples are montmorillonite and bauxite).

## CLEAVAGE

As explained previously, a crystal is made up of atoms which are arranged into atomic units that join together into a structure which is basically self determining. The whole framework into which these atomic units fit is a three-dimensional network of atomic bonds; these collectively form a crystal lattice.

These bonds do not necessarily possess the same strength. When subjected to mechanical stress, a crystal will tend to break cleanly along a lattice plane if the atomic bonds are strong in all but one direction. This crystal will have perfect cleavage. When the crystal does not show a dominant direction of weak or strong bonds, it will not break cleanly, and is said to display poorly defined cleavage.

The presence of a cleavage plane is an inherent part of a crystal's composition. It is a characteristic that is often used when working with hard and precious stones to facilitate the process of shaping them.

## SPECIFIC GRAVITY

The specific gravity of a crystal (sometimes referred to as its relative density) is defined as the weight per unit volume, ie the ratio of the weight of an object to that of an equal volume of distilled water. Consequently, a crystal with a specific gravity of 4 is four times as heavy as the same volume of water.

The specific gravity of a crystal can be measured in a number of ways. Firstly, using a hydrostatic balance, the crystal is weighed in air (W1) and then in distilled water (W2). Its specific gravity can then be calculated as: **specific gravity** $= \dfrac{\textbf{W1}}{\textbf{W1} - \textbf{W2}}$

Diagram of a hydrostatic balance

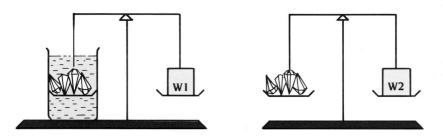

An alternative method uses a pycnometer ('density measurer'). This device is a bottle with a tightly fitting stopper which has a capillary hole in it. The procedure adopted consists of firstly weighing the crystal on scales (see diagram, W1). The pycnometer is then filled with distilled water, and, with the stopper in place, is also weighed (W2). Next, the crystal is placed in the pycnometer and the stopper replaced. The excess water is now removed, and the pycnometer and its contents are weighed once more (W3). The specific gravity of the crystal can now be calculated using the formula: **specific gravity** $= \dfrac{\textbf{W1}}{\textbf{W1} + \textbf{W2} - \textbf{W3}}$

Diagram of the three stages of using a pycnometer

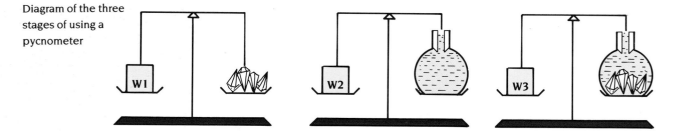

The density of a material is a function of how closely together the atoms are packed. It is often used by the mineral processing engineer as an easy and reliable way of separating a useful mineral from a worthless one, especially if there is a large differential in their respective specific gravities.

## CRYSTAL USES, PAST AND PRESENT

Crystals were originally used by primitive man as a cutting edge; they were attached to a wooden shaft or handle, which could be used as a formidable weapon or tool. They were also used by cave dwellers in the paintings they made on the sides of rocks and cave walls, some of which still exist in southern France, northern Spain and eastern Australia. These paints were made from ground and powdered hematite (red), limonite (yellow) and pyrolusite (blue-black) which were mixed with water or animal fats to produce a paste. They have also been used in this form as a body paint to be worn in battle, at ceremonies or as a religious decoration.

Over 5,000 years ago in Ancient Egypt, the Pharaohs employed thousands of slaves to work the turquoise mines located within the Sinai Peninsula. These were perhaps the first commercially operated crystal mines. In addition to their quest for turquoise, the Egyptians had an almost fanatical desire for lapis lazuli. As the demand for this beautiful crystal could not be met by local production, supplies were carried from Afghanistan, some 2,500 miles away. Imitations were also produced to meet deficiencies in supply.

*ABOVE Crystals were often carried into battle, as with the garnets on this silver belt buckle found in a 6th century Viking grave in Norway.*

*LEFT Long sought after for its beauty and endurance, gold is a rare metal which may be found in crystal form. This gold dredge is working a river bed on New Zealand's South Island.*

*BELOW This wooden box contains a selection of typical Egyptian jewellery and amulets, featuring such crystals as gold and lapis lazuli.*

Nowadays some of the most commonplace materials, such as plaster of Paris and rock salt, come from crystal sources. Crystals are used in the metals industry not only as a source of most metal ores, but as the main ingredient of refractory bricks and the fluxes used in metallurgical refining. High quality crystals continue to be used as gemstones.

A revolutionary step in crystal engineering has recently come about as a direct consequence of the interest shown in ceramics as an alternative material to steel alloys. Ceramic materials and powdered metal are pressed into a mould at such high temperatures and pressures that they fuse into a solid mass. The resulting mass is machined to precise tolerances and may be used as, say, an engine block. The reduction in weight and the degree to which the original powdered mass can be sculptured have been welcomed as a breakthrough in materials engineering. The next stage in this exciting development was the growing of a single crystal into the shape of an engine block or turbine blade using a heavily charged solution as the source medium. Work is continuing in this field and encouraging progress has already been made.

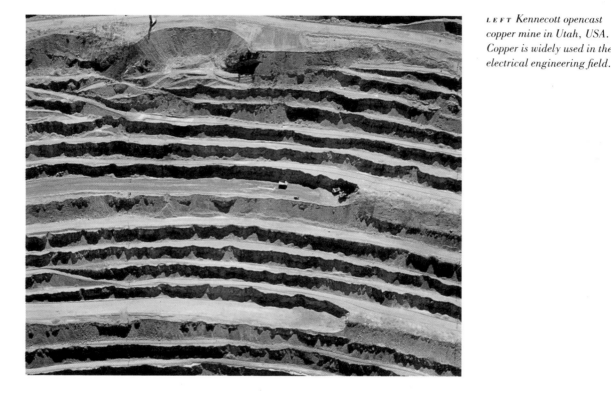

*LEFT Kennecott opencast copper mine in Utah, USA. Copper is widely used in the electrical engineering field.*

## SYNTHETIC CRYSTALS

The Egyptians were the first people to produce imitation stones. They coloured alkali stones with blue copper cobalt to produce a stone similar in appearance to lapis lazuli. Not all imitations were designed to deceive; today, synthetic crystals are an essential part of modern industrial practices.

In 1758 Joseph Strasser, a Viennese scientist, produced a type of glass which could be cut to look like a diamond. However, the major step forward in synthetic crystallization occurred in 1902 when A. V. L. Verneuil, a French chemist, produced a synthetic ruby by accident – whilst he was trying to produce a sapphire!

The method that Verneuil employed is now known as the flame-fusion process. Powdered raw material is heated in a furnace to 3,632°F (2,000°C), the molten material drips through a hole and is collected to form a short candle-shaped mass called a 'boule'. It takes about four hours to form a complete boule which is then split lengthways to relieve any internal stresses. The boule is then ready to be cut into a crystal. This method is also used to produce other gemstones such as sapphires, aquamarines, blue zircons, emeralds, and tourmalines.

A synthetic diamond was first produced in 1955 when a graphite crystal was subjected to such extremes of temperature and pressure that the carbon atoms rearranged themselves into a diamond. After the initial breakthrough, development work continued until a crystal of gem quality was produced in 1970. However, this process is extremely expensive, since the conditions which existed below the earth's crust when natural diamonds were first formed, have to be duplicated in a laboratory. This method requires the generation of pressure between 50,000 and 100,000 pounds per square inch (344,750 – 689,500 kilopascals), at a temperature of 2,732 to 4,352°F (1,500 to 2,400°C). Despite the cost, synthetic diamonds which are not of gem quality are still being produced and play an important part in industrial cutting, drilling and grinding.

Synthetic crystals can be produced to a consistently high standard, and this has led – despite the availability of the natural article – to the transformation of the synthetic crystal industry from an interesting laboratory process into an immensely lucrative business.

Crystals are truly remarkable objects; usually formed one atom at a time, within a heated and highly pressurized environment, it is astonishing that they exist at all. Irrespective of their appearance – perfect or imperfect, bright and beautiful or dull and ugly – it is impossible to imagine a world without them.

*ABOVE A scene from the De Beers diamond sorting house.*

*LEFT Brownish yellow synthetic diamonds manufactured at the De Beers Diamond Research Laboratory.*

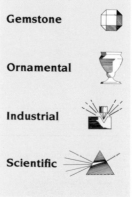

NOTES ON USING THE CRYSTAL IDENTIFIER

*The crystals in the identifier section are listed in descending order of hardness, using Mohs' scale. To find the reference for a specific crystal name, refer to the index at the back of the book.*
*Below each identifier entry is a group of symbols that indicate the hardness, specific gravity, crystal system (or classification) and main use of the crystal described. The symbols to indicate the use of the crystal are as follows:*

**Gemstone**

**Ornamental**

**Industrial**

**Scientific**

# Crystal
# Identifier

# Diamond

The word 'diamond' comes from the Greek 'Adamas', meaning invincible. Its value as a gemstone is based on the 'four Cs'; colour, clarity, cut and carat (weight). Only 20 per cent of all diamonds are suitable for gemstones, with the rest being used in industry as a high-quality abrasive for drilling, boring and grinding. Diamonds are formed at great depths by high temperature and pressure; then they are blasted by volcanic forces through vents towards the surface at such a speed that they are unable to cool into graphite, but form diamonds instead.

Some diamonds are colourless; others come in different tones of yellow – light, dull or brown tinted. In rare cases diamonds display a strong colour (blue, brown, green, red, violet or yellow) and are referred to as 'fancy coloured'. A diamond's most distinctive features are its hardness and sparkle, which are unmatched by any other crystal.

**Range:** South Africa is the world's major supplier of gem-quality crystals, with the majority of industrial diamonds coming from deposits in Angola, Australia and Zaire.

---

### DIAMOND

**T**he diamond is considered the stone of the mind because of its hardness. It can be used to stimulate clear thoughts or prevent dreaming. The legendary brilliance of the stone is reputed to form a barrier against negative thoughts. Eros, the Greek god of love, tipped his arrows with diamonds to ensure that those whose hearts they struck would fall in love with each other. The diamond is the gemstone of the Libra star sign.

---

### RUBY

**T**he ruby is often considered to be the prince among crystals as it is said to contain the heritage of humanity, the gift of humility and the beauty of a pure spirit. Rubies help the body to survive in times of peril and are often associated with authority and natural leadership. Over the centuries, the ruby crystal has become a symbol for everlasting love and loyalty. In ancient times it was believed that the stone contained an inner heat which had the power to boil water. Rubies are associated with the Leo star sign.

---

*TOP LEFT A group of uncut natural diamonds from South Africa, showing the colour variation possible in these gemstones.*

*TOP RIGHT Ruby crystals in a quartz matrix; mined in Pakistan.*

# Ruby

*CORUNDUM GROUP*

Rubies have long been prized as a precious gemstone. Lower quality crystals are powdered and used as a high-quality cutting and polishing medium. Synthetic rubies have replaced the 'jewel' rubies which were traditionally used as bearings in delicate instruments. They tend to occur in dolomitic type limestones which have become marble-like rocks. Most rubies are mined from alluvial deposits.

Rubies occur in a variety of shades of red with chrome being the colouring pigment; any brown hue which may occur is due to the presence of iron.

The distribution of colour within the crystal is often uneven, occurring in strips or spots; different tones of red can often occur within crystals from a single deposit. Ruby crystals become distinctly darker in colour when they are exposed to natural light. They often possess a soft, silky sheen which is due to the inclusion of minute rutile crystals.

**Range:** The highest quality rubies come from Burma, although Thailand supplies most of the world's markets. Small quantities of rubies also come from Afghanistan, Australia, Brazil, Cambodia, India, Malawi, Pakistan and the United States (Montana and North Carolina).

---

**Mohs' Hardness** *10*
**Specific Gravity** *3.47 – 3.55*
**Crystal Structure Isometric**
**Gemstone, Industrial**

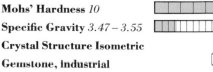

**Mohs' Hardness** *9*
**Specific Gravity** *3.97 – 4.05*
**Crystal Structure Hexagonal**
**Gemstone, industrial**

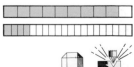

# Sapphire

### CORUNDUM GROUP

A sister crystal to ruby, the comparatively lower value of the sapphire reflects the relatively common occurrence of iron within the stone, from which it derives its distinctive blue colour. Sapphires are formed as crystals in marble, basalt and pegmatite, as a result of contact metamorphism between alumina-rich magma-type rocks and limestone.

Sapphire crystals occur in a variety of colours; black, purple, violet, dark blue, bright blue, light blue, green, yellow and orange. Traces of titanium contribute along with the presence of iron to blue colours; violet colours come from vanadium being present, and any red hues from the presence of chrome.

The colour of the stone is important in determining the point of origin of the crystal: Australian stones have a deep blue colour with distinctive blue-green pleochroism; Sri Lankan stones have a patchy blue colour, while Kashmir stones are a deep milky cornflower blue with no obvious inclusions.

**Range:** The most desirable crystals traditionally come from the Kashmir region of India but Australia is now the world's foremost source. Sapphires are also found in Brazil, Burma, Cambodia, Kenya, Malawi, Tanzania, Thailand and Zimbabwe.

SAPPHIRE

*The sapphire is known as the jewel of truth and wisdom and is associated with heavenly inspiration, devotion and spiritual control. The crystal is also reputed to be capable of controlling desire and passion. Some cultures believe that the three points contained within the star sapphire represent destiny, faith, and hope; consequently these stones are considered to be particularly useful for transforming wishes into reality.*

*TOP LEFT A Sri Lankan sapphire with the distinctively patchy blue colouring of sapphires from that region.*

*TOP RIGHT A sample of green chrysoberyl, a relatively rare crystal.*

# Chrysoberyl

Green chrysoberyl, alexandrite (named after its discovery in the Urals in 1830 on the day Prince Alexander of Russia came of age) and honey-yellow cat's-eye, are all highly prized gemstones. Although the crystal is uncommon, when it does occur, it is usually associated with granite-pegmatites, mica schists, alluvial or marine deposits.

Chrysoberyls can be green, yellow, grey, brown, or colourless. Alexandrite is very distinctive in that it is green in natural light and red in artificial light; Green grossular garnets, which have recently been discovered in East Africa, display similar characteristics.

Cat's-eyes are greenish yellow or yellow, often with a cold greyish tone; they display a moving light ray which along with their colour lead them to resemble the eyes of feline creatures. Chrysoberyls tend to be sensitive to chemical attack by alkalis and can change colour when heated.

**Range:** Chrysoberyls are found in Brazil, Burma, Madagascar, Norway, Sri Lanka, Tanzania, the USSR and the United States (Connecticut) and Zimbabwe.

**Mohs' Hardness** *9*
**Specific Gravity** *3.99 – 4.00*
**Crystal Structure Hexagonal**
**Gemstone, Industrial**

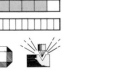

**Mohs' Hardness** *8½*
**Specific Gravity** *3.7 – 3.72*
**Crystal Structure Orthorhombic**
**Gemstone**

# Spinel

The spinel crystal has been used by jewellers for centuries, with quality cut gems being set side-by-side with some of the most famous diamonds and rubies in the world. Today the market for spinel has become somewhat confused because of the creation of good quality synthetic stones; these were a by-product of early attempts to create synthetic sapphires. Spinel crystals are usually formed in limestones by regional or contact metamorphism, especially when there is an abundance of manganese or aluminium.

Spinel crystals come in a variety of colours including pink, red, violet red, pale lilac, blue, violet blue and black. The red hue is due to iron being present in the stone, while the blue hue is due to the presence of chrome. Crystals can be transparent or opaque with a vitreous lustre.

**Range:** Gem-quality red and pink spinel predominantly come from Afghanistan and Burma where often they are found with rubies. Blue sapphires, considered to be the most valuable, come from the gem gravels of Sri Lanka. Smaller deposits occur in Brazil, Madagascar, Thailand and the United States (New Jersey and New York State).

## TOPAZ

**T**opaz means 'fire' in Hindu, and in living up to its name brings light to life. The stone is believed to relieve stress; a crystal placed under the bed at night is reputed to have revitalizing properties, stimulating refreshing dreams. It is reputedly beneficial for the lungs and is therefore useful in preventing colds. Topaz is the gemstone of the star sign Scorpio.

*TOP LEFT Gem-quality red spinel from the Mogok region of Burma, one of the world's main sources of fine spinel.*

*TOP RIGHT Good examples of gem-quality blue topaz from Australia.*

# Topaz *(or Precious Topaz)*

Topaz is a highly prized gemstone. In recent years, however, it has lost much of its popularity to citrine quartz, which is similar in appearance. Pinkish-yellow topaz from Brazil is often heat-treated to give pink topaz, while colourless topaz becomes blue when irradiated. Topaz crystals are formed by the crystallization of magmatic fluid under great pressure. They are mostly found in alluvial deposits. Despite their hardness, they are susceptible to chemical weathering, especially in tropical and equatorial regions.

There are four varieties of topaz which are categorized by colour: true (or yellow) topaz, pink topaz, blue topaz and colourless topaz. The most common crystals have a yellow colour with a red tint. Deep-yellow crystals are the most valuable.

**Range:** Brazil is the world's major source of gem quality topaz. Significant deposits also occur in Afghanistan, Burma, Japan, Mexico, Namibia, Nigeria, the USSR, the United States, Zambia and Zimbabwe.

**Mohs' Hardness** *8*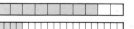
**Specific Gravity** *3.58 – 3.61*
**Crystal Structure Isometric**
**Gemstone**

**Mohs' Hardness** *8*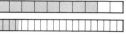
**Specific Gravity** *3.52 – 3.56*
**Crystal Structure Orthorhombic**
**Gemstone**

# Aquamarine

*BERYL GROUP*

The name aquamarine derives from Latin and means 'water from the sea'. It was traditionally a good luck charm favoured by sailors. Rich blue crystals are the most highly prized as gems, while pale and green stones are still used extensively for necklaces and brooches. Although aquamarine crystals are associated with magmatic intrusions they are usually found in alluvial deposits, where their hardness and resistance to chemical attack have allowed them to remain largely unaltered.

Aquamarines vary in colour from pale blue to light blue-green to light green. Green crystals, when they occur, are watery with no hint of yellow; this is due to iron coloration. Light blue crystals can be heated to 752°F (400°C) to alter them to a deeper blue hue. Aquamarines are distinguishable from topaz stones and spinels by their stronger lustre and lack of grey or violet tint.

**Range:** Aquamarines occur on all continents, with the most substantial gem-quality stones coming from Afghanistan, Brazil, Madagascar, the USSR and the United States.

AQUAMARINE

**A**quamarines are renowned for their tranquil qualities and their ability to bring peace and calm to the most troubled of minds. The crystals have a cooling power and as such can be used to calm a fever or relax an overactive mind. Aquamarines are thought to cure motion sickness as well as cleanse glands, relieve water retention and reduce nervous disorders. Aquamarine is the gemstone of the Gemini star sign.

*TOP LEFT An interesting example of a light green aquamarine crystal from Pakistan.*

*TOP RIGHT A very pale green variety of beryl; this crystal was found in Africa.*

# Beryl *(or Precious Beryl)*

*BERYL GROUP*

The German word 'brille', meaning eye-glasses, is derived from the word beryl, since eye lenses were once ground from colourless beryl. Beryls are characteristic of granitic and pegmatitic rocks where they often occur as large crystals. They are also formed by chemical substitution in a metamorphic environment, as well as being formed from the crystallization of hydrothermal fluids. Due to their hardness and resistance to chemical attack, beryls often occur as unaltered crystals in alluvial deposits.

Beryl, which is considered a semi-precious gemstone, refers to all crystals of the beryl group which are not emerald-green or aquamarine-blue. Beryl, although hard, is brittle and easily split along ill-defined planes. Beryl crystals, which have a vitreous lustre, are resistant to all chemicals with the exception of fluoric acid.

**Range:** Beryl occurs throughout the world. Important deposits occur in Brazil, Czechoslovakia, India, Madagascar, the USSR and the United States (Connecticut, Maine, New Hampshire and South Dakota).

**Mohs' Hardness** *7½ – 8*
**Specific Gravity** *2.67 – 2.71*
**Crystal Structure Hexagonal**
**Gemstone**

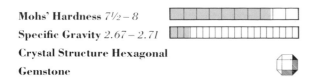

**Mohs' Hardness** *7½ – 8*
**Specific Gravity** *2.65 – 2.75*
**Crystal Structure Hexagonal**
**Gemstone**

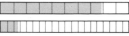

## Bixbite *(or Red Beryl)*

*BERYL GROUP*

Due to its rarity, distinctive colour and the publicity surrounding its recent discovery, bixbite is a relatively highly valued semi-precious gemstone. While commanding a high market price, bixbite has not as yet been commercially imitated or produced synthetically. Unlike the other crystals of the beryl group which are found in or near pegmatic veins, bixbites are found in effusive magmatic rhyolite rocks.

Bixbite has a strong, ruby-red, violet or strawberry-red hue. The crystals, which tend to be small, always contain numerous inclusions and more often than not internal flaws.

**Range:** Bixbite crystals are only found in the United States (New Mexico and Utah).

*TOP LEFT This very dark example of bixbite crystal is from Utah, USA.*

*TOP RIGHT Emerald crystals from pegmatite veins, mined in Colombia.*

## Emerald

*BERYL GROUP*

The emerald has long been valued as a precious gem, those with deep-green crystals commanding the highest prices. Synthetic emeralds are now being produced with artificial inclusions similar to those found in natural crystals. Emeralds are formed in the vicinity of rising magma and are characteristic of granites and pegmatites where they occur as crystals.

Emerald crystals are bright green, light green, yellow-green or dark green. The pigment is due to traces of chrome and sometimes vanadium. The colour is very stable in light and heat, only altering at 1,292 – 1,472°F (700 – 800°C). Emeralds are only occasionally transparent; they mostly contain inclusions which are due to liquid or gas bubbles, heating cracks, or foreign crystals. These inclusions are considered to be a sign of authenticity and do not necessarily detract from the value of the crystal.

**Range:** The most important emerald source is Colombia, where the crystal – which occurs in calcite veins – was originally mined by the Incas. Emeralds also occur in deposits in Australia, Brazil, India, Pakistan, South Africa, Tanzania, the USSR, Zambia and Zimbabwe.

**Mohs' Hardness** *7½ – 8*

**Specific Gravity** *2.65 – 2.75*

**Crystal Structure Hexagonal**

**Gemstone**

**Mohs' Hardness** *7½ – 8*

**Specific Gravity** *2.67 – 2.78*

**Crystal Structure Hexagonal**

**Gemstone**

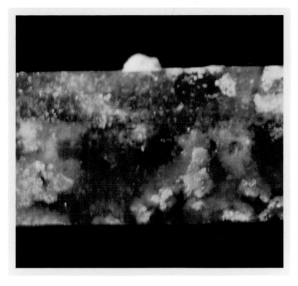

# Golden Beryl

### BERYL GROUP

Golden beryl is one of the less valuable gems from the beryl group, although brightly coloured specimens are always in demand as gemstones. As it is of marginal value, it is not produced synthetically or imitated. As with most of the other crystals from the beryl group, golden beryl is associated with igneous activity.

The most highly prized golden beryl crystals range in colour from canary yellow to golden yellow, while other stones are dull yellow or lemon yellow. The pigment is thought to be due to iron contained within the crystal structure. Although inclusions are rare, when they do occur, they take the form of regular parallel bundles in a straw-like form, which – clearly visible with a lens – reduce the crystal's transparency and lustre.

**Range:** Golden beryl comes primarily from the Minas Gerais state of Brazil, although other deposits occur in Madagascar, Sri Lanka and the United States (Massachusetts and Virginia).

*TOP LEFT This attractively coloured golden beryl crystal is from Brazil.*

*TOP RIGHT A rose coloured sample of morganite from Utah, USA.*

# Morganite *(or Pink Beryl)*

### BERYL GROUP

Morganite, which is named after the American banker and mineral collector John Morgan (1837–1913), is a strongly coloured crystal which is much valued as a gemstone. Morganite crystals occur in or near pegmatite veins.

Morganite crystals are usually soft pink or violet in colour, with no overtones. Lower colour quality specimens can be heat-treated to 752–932°F (400–500°C) where they improve into aquamarines. The crystals, which usually occur as long prisms, have a glassy lustre. Although Morganite crystals are usually free of inclusions, when they do occur, they are very irregularly shaped liquid or gaseous forms, which are only just visible.

**Range:** Important morganite deposits occur in the Minas Gerias region of Brazil, Madagascar, Mozambique, Namibia, the United States (California) and Zimbabwe.

**Mohs' Hardness** $7\frac{1}{2} - 8$
**Specific Gravity** $2.65 - 2.75$
**Crystal Structure Hexagonal**
**Gemstone**

**Mohs' Hardness** $7\frac{1}{2} - 8$
**Specific Gravity** $2.8 - 2.9$
**Crystal Structure Hexagonal**
**Gemstone**

# Phenacite *(or Phenakite)*

Quality transparent phenacite crystals are sometimes cut into gemstones. Phenacite has been produced synthetically on a limited basis. The crystals are formed in high-temperature pegmatite veins and in mica schists. They are frequently found with quartz, chrysoberyl, beryl, apatite and topaz.

Phenacite crystals are colourless, white, yellow tinted or pink. Crystals are transparent with a vitreous lustre. The crystal's cleavage is imperfect, but it is infusible and insoluble in acids.

**Range:** Phenacite crystals are found in Brazil, Mexico, the USSR (the Urals) and the United States (Colorado).

*TOP LEFT A high-quality sample of the rare crystal, phenacite, from the USSR.*

*TOP RIGHT Andalusite crystals lying on a matrix of pelagic crystals, from the UK.*

# Andalusite

When available in large enough quantities, andalusite is used in industry for the manufacture of refractories, high-temperature electrical insulators, and acid-resistant ceramic products. Quality crystals are sometimes cut into gems, especially when they display a greenish or reddish colour. Andalusite is a characteristic mineral of low-pressure metamorphic rocks (usually granitic or argillaceous). They are rich in aluminium and poor in calcium, potassium and sodium.

The colour of andalusite, varies from light-yellowish brown to green-brown, light brownish-pink, bottle green or greyish-green. Andalusite crystals have a modest lustre and sometimes have dark inclusions running across the plane of the prism; these crystals are called chiastolite.

**Range:** Fine andalusite crystals are found in Spain in the vicinity of Andalusia (hence the name), as well as in Australia, Brazil, Burma, Canada (Quebec), Sri Lanka and the United States (California, Maine, Massachusetts and Pennsylvania).

**Mohs' Hardness** *7½ – 8*

**Specific Gravity** *2.95 – 3.0*

**Crystal Structure Hexagonal**

**Gemstone**

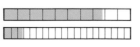

**Mohs' Hardness** *7 – 7½*

**Specific Gravity** *3.12 – 3.2*

**Crystal Structure Orthorhombic**

**Industrial, Gemstone**

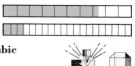

# Danburite

Danburite is of marginal value as a gemstone, despite its hardness. The crystals are usually found in fissures and in lining lithoclase, especially as an incrustation in albite.

Danburite crystals tend to be either colourless or pink; occasionally they are pale yellow. The crystals have poor cleavage and are transparent with a greasy lustre. The boron in the crystal's make-up will colour a flame green, and it can easily be fused into a colourless glass. The danburite crystal is insoluble in acid.

**Range:** It is named after Danbury (Connecticut, United States), where the crystal was first found. Other deposits occur in the Swiss and Italian Alps, Japan, Madagascar and Mexico.

*TOP LEFT A pale pink example of danburite from Mexico.*

*TOP RIGHT A deep blue/violet sample of Brazilian indicolite.*

# Indicolite

*TOURMALINE GROUP*

Attractive clear blue, bright blue or blue-green crystals of indicolite are often cut and set as gemstones. However, when the colour is too deep or too blue, the stones fail to attract much interest from jewellers. Indicolite crystals commonly occur in greisen and pegmatites, where they have grown due to magmatic intrusions. Crystals also occur in sedimentary rocks as branch-like or authigenic grains.

Indicolite, named after the colour indigo, is the blue variety of tourmaline, but crystals can also be a greenish blue colour. The colour is a distinctive feature of indicolite, as is the loss of transparency when viewed in one direction. The crystals are insoluble in acid.

**Range:** Indicolite crystals are found in Brazil, Madagascar, Namibia, the USSR (the Urals) and the United States (California, Colorada and Massachusetts).

**Mohs' Hardness** *7 – 7½*
**Specific Gravity** *2.97 – 3.02*
**Crystal Structure Orthorhombic**
**Gemstone**

**Mohs' Hardness** *7 – 7½*
**Specific Gravity** *3.02 – 3.26*
**Crystal Structure Hexagonal**
**Gemstone**

# Iolite *(or Cordierite or Dichroite)*

The blue variety of iolite, which is known as water sapphire, is cut into a moderately valuable gemstone. Iolite is formed in areas of contact metamorphism and less frequently in silica- and alumina-rich granitic or rhyolitic eruptive rocks.

The name itself is derived from its sometimes violet colour. The crystals are normally a variety of dark blue to light blue, with grey being the most common additional hue. There is also a black, iron-bearing iolite called sekaninaite. The crystal has sometimes been referred to as the 'Vikings' compass' in reference to its ability to indicate the direction of the sun on overcast days. It is sometimes confused with quartz; unlike quartz, however, it is fusible in thin sections and is insoluble in acid.

**Range:** Iolite crystals are mainly found in Brazil, Burma, Canada, Finland, India, Madagascar, Sri Lanka, Namibia and the United States (Connecticut).

*TOP LEFT Iolite, also known as cordierite or dichroite, often takes on a violet-grey hue, as in this sample from Brazil.*

*TOP RIGHT Gem-quality pyrope crystals from alluvial deposits in Czechoslovakia.*

# Pyrope *(or Cape Ruby)*
### GARNET GROUP

Pyrope gets its name from the Greek word *pyropos*, meaning fiery. It is a moderately valuable semi-precious gemstone, with the darkest crystals being the most common. Pyrope is typically formed in peridotite and serpentinized peridotites, as well as in the diamond-bearing clay called kimberlite. Due to their resistance to weathering, pyropes are often found in alluvial or secondary deposits.

Pyrope is deep red in colour, with the coloration due to traces of chrome in the crystal structure. Pyrope is singly refractive with birefrigence patches. Its lustre is comparable to that of rubies and spinels. Pyrope fuses fairly easily and is almost insoluble in acid.

**Range:** Pyrope crystals are found in moderately sized deposits in Argentina, Australia, Brazil, Czechoslovakia, Mexico, South Africa, Tanzania and the United States.

**Mohs' Hardness** *7 – 7½*

**Specific Gravity** *2.53 – 2.66*

**Crystal Structure Orthorhombic**

**Gemstone**

**Mohs' Hardness** *7 – 7½*

**Specific Gravity** *3.65 – 3.87*

**Crystal Structure Isometric**

**Gemstone**

## Rubellite

### TOURMALINE GROUP

Rubellite crystals which are ruby-coloured and contain no inclusions are modestly priced and used as gemstones. Rubellites are usually formed in association with igneous and metamorphic activity. Crystals commonly occur in greisen and pegmatites but can also occur in sedimentary rocks as detrital and authigenic grains.

The colour of rubellite crystals varies from pink to violet pink, pink with a brown tint, to red or violet-red with a brownish tint. Violet-red rubellite crystals are sometimes referred to as siberite, after Siberia where they occur. The colour is a distinctive feature of rubellite crystals. However, sometimes the colour can be a little subdued. The crystals can also fail to become brighter in strong light, something they have in common with rubies.

**Range:** Rubellites are found in Brazil, Burma, Madagascar, Sri Lanka, the USSR (Siberia) and the United States (California).

*TOP LEFT A rose-pink example of gem-quality rubellite mined in Brazil.*

*TOP RIGHT An elongated crystal of red staurolite, shown here with pale blue kyanite; this sample is from Switzerland.*

## Staurolite *(or Fairy Stones)*

Staurolite (the word derives from the Greek *stauros* and *lithos*, meaning stone and cross) frequently occurs in cruciform-twinned crystals; the crystals either form a Greek cross (90 degrees between the arms) or a St Andrew's cross (60 degrees between the arms). Subsequently, they are often worn as religious jewellery. Staurolite is a metamorphic mineral formed within a medium-temperature environment.

Staurolite crystals often have a rough earthy coating (due to surface chemical alteration) which hides their reddish-brown to black interior. They frequently occur embedded in kyanite or schist as flat elongated crystals. Crystals are generally semi-opaque, with a resinous lustre. They are infusible but slightly soluble in sulphuric acid.

**Range:** Staurolite crystals are found in East Germany, Scotland and Switzerland. The cruciform-twinned crystals regularly occur in France and the United States (Georgia, New Mexico and Virginia).

**Mohs' Hardness** *7 – 7½*
**Specific Gravity** *3.02 – 3.26*
**Crystal Structure Hexagonal**
**Gemstone**

**Mohs' Hardness** *7 – 7½*
**Specific Gravity** *3.7 – 3.8*
**Crystal Structure Monoclinic**
**Gemstone**

# Tourmaline

Tourmaline was named from the Sinhalese word 'tora-malli' meaning high or hard rocks. Tourmaline will develop a positive and negative charge when rubbed or heated. High quality, consistent tourmaline crystals are cut into gemstones; dramatically zoned stones often are sliced and polished as ornamental objects.

Tourmalines either occur in intrusive dikes or silica-rich rock (especially granites and pegmatites), or as well defined crystals due to their hardness.

Tourmaline occurs in a wide variety of colours; the most attractive are pink, fiery red and deep green. Pale coloured crystals can be heat-treated at 842 – 1,202°F (450 – 650°C) to intensify coloration. Colour zoning is common, and can vary from green at the base of a crystal to red at the apex. Crystals are brittle and are commonly elongated or columnar.

**Range:** The most widely occurring tourmaline is the black, iron-rich variety known as schorlite which is of little value. Sizeable and highly utilized deposits of other varieties of tourmaline occur in Brazil, Burma, Sri Lanka, Madagascar, Mozambique, the USSR (the Urals) and the United States (California, Connecticut and Maine).

*TOP LEFT This pink and green tourmaline crystal demonstrates the colour zoning common in this type of crystal.*

*TOP RIGHT Verdelite, or green tourmaline, shown here in elongated crystals.*

# Verdelite *(or Green Tourmaline)*

### TOURMALINE GROUP

The name verdelite is derived from the Italian word meaning 'green stone'. Verdelite crystals range in colour from light-green to emerald-green to deep-green. Only clear crystals from the mid-range of colours are considered suitable for cutting into gems.

The most distinctive features of verdelite crystals are their dark shades and the loss of transparency along the crystal axis. As with other crystals from the tourmaline group, verdelite crystals are associated with igneous, intrusive and metamorphic environments, although they are commonly recovered from alluvial deposits.

**Range:** Verdelite crystals are found in Brazil, Mozambique, Namibia, Sri Lanka, Tanzania, the USSR and the United States (Maine).

**Mohs' Hardness** *7 – 7½*
**Specific Gravity** *3.02 – 3.26*
**Crystal Structure Hexagonal**
**Gemstone, Ornamental**

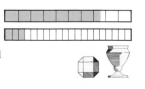

**Mohs' Hardness** *7 – 7½*
**Specific Gravity** *3.02 – 3.26*
**Crystal Structure Hexagonal**
**Gemstone**

## Amethyst

QUARTZ GROUP

The name Amethyst is derived from the Greek word *amethystos*, meaning 'not drunk', as it was believed that by wearing the crystal one was protected from the effects of alcohol. The Amethyst is the most highly prized form of quartz. However, its value fell at the turn of the century with the discovery of large deposits in Brazil and Uruguay. High-quality crystals are still often used as a semi-precious gem. Amethyst crystals always grow from a base.

The pyramid-like crystals are not usually well developed and are therefore frequently found with the deepest colour (violet, purple or pink) at the tips. The colour is due to traces of ferric iron. The distribution of varying colour bands distinguishes amethysts from crystals of similar appearance. The presence of broken, wavy parallel lines helps to make them distinguishable also; these are due to inclusions. A crystal will turn white when heated to 572°F (300°C), and will yellow at 932°F (500°C).

**Range:** Some spectacular crystals come from the state of Minas Gerais in Brazil. Amethysts are also found in large quantities in Australia, Canada, Czechoslovakia, India, Madagascar, South Africa, Sri Lanka, the USSR and the United States.

### AMETHYST

**T**he amethyst crystal is thought to have the power to transmit calm. It is also believed to uplift the mind and to improve general well-being, as well as alleviating insomnia. It has a powerful healing influence and can generate spiritual awakening. It can also be used as an aid to chastity through its ability to calm strong emotions.

### CITRINE

**C**itrine is a sun stone; it enables one to develop clear lines of thought, direct arguments and precise communication skills. Citrine crystals are reputed to help the body rid itself of physical and emotional problems.

*TOP LEFT These Brazilian amethysts have grown inside a geode, a common place for these crystals to form.*

*TOP RIGHT This citrine crystal is a heat-treated amethyst, mined in Brazil.*

## Citrine

QUARTZ GROUP

Most commercial citrines are actually heat-treated amethysts or smoky quartz. Citrine is widely used as an imitation of the more expensive gemstone, topaz. The larger citrine crystals, which are prismatic with pyramid ends, are associated with intrusive magmatic phenomena.

The distinctive colour is due to the presence of colloidal iron hydrates and varies from pure yellow to dull yellow, honey or brownish yellow. Citrine crystals will turn white if heated and dark brown if exposed to X-rays. As with amethysts, the colour is often broken up into patches or bands. Cut crystals display good lustre. The density of citrine is the lowest for crystals of this colour.

**Range:** Citrine is rare but does occur in Brazil, France, Madagascar, the USSR and the United States (Colorado).

**Mohs' Hardness** 7
**Specific Gravity** 2.63 – 2.65
**Crystal Structure Hexagonal**
**Gemstone, Ornamental**

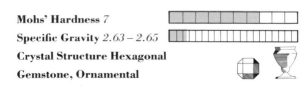

**Mohs' Hardness** 7
**Specific Gravity** 2.65
**Crystal Structure Hexagonal**
**Gemstone**

# Milky Quartz

QUARTZ GROUP

Milky quartz is often cut into beads, ornaments and *objets d'art*. The most common variety of milky quartz is found in pegmatites and hydrothermal veins; mystery still surrounds the process which leads to their formation and concentration.

The distinctive coloration of milky quartz is due to the inclusion of numerous bubbles of gas and liquid in the crystal. Milky quartz is the rougher, more compact formation of amethyst, layered and striped with milky bands.

**Range:** Milky quartz is one of the most common materials in the earth's surface (12 per cent by volume), and occurs in many locations. Famous finds include a 14.5 ton (13 tonne) crystal in the USSR (Siberia). It is especially common in central Europe, Brazil, Madagascar, Namibia and the United States.

*TOP LEFT A dramatic formation of milky quartz from the USA, showing clearly the hexagonal form of this crystal type.*

*TOP RIGHT Although common, rock crystals are still popular as gemstones or for ornamental use; this example is from Switzerland.*

# Rock Crystal
## *(or Colourless Quartz)*

QUARTZ GROUP

The name quartz comes from the Greek for ice, since it was once believed that the crystals were forever frozen by a process of extreme cold. Although quite common, rock crystal is often carved into *objets d'art* or fashioned into jewellery. In the past it has been used for optical and piezoelectrical purposes; synthetic crystals are now generally used for this purpose. Rock crystal is crystallized directly from magma, in pegmatites, and in low-temperature hydrothermal regions.

Rock crystal is colourless, transparent, and – unlike glass – is birefringent. It is distinguishable from ordinary glass by its absence of air bubbles, and from lead-glass by its hardness (7 compared to 5).

**Range:** Quartz is one of the most commonly occurring minerals in the earth's crust (12 per cent by volume). Brazil has produced some spectacular crystals which have weighed in excess of 4 tons (3.6 tonnes).

**Mohs' Hardness** *7*
**Specific Gravity** *2.65*
**Crystal Structure Hexagonal**
**Gemstone, Ornamental**

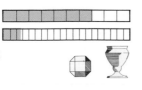

**Mohs' Hardness** *7*
**Specific Gravity** *2.65*
**Crystal Structure Trigonal**
**Gemstone, Ornamental**

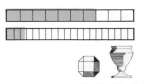

# Rose Quartz

QUARTZ GROUP

Rose quartz is much valued as an ornamental material because of its attractive colour and comparative rarity. However, it is not very popular because of its tendency to be brittle. Although rose quartz occurs in massive form in many pegmatites, well formed crystals are rare.

The colour varies from strong to pale pink, and appears to be caused by traces of manganese or titanium. The crystal is usually somewhat milky rather than perfectly transparent. Named after its colour, the crystal is often cracked, and usually a little turbid. It is only in recent years that crystals with flat sides have been found. Crystals tend to lose their colour when heated; they turn black when exposed to radiation.

**Range:** Rose quartz is not uncommon, but is usually found in compact masses. Quality crystals have been found in Brazil, Madagascar and the United States (California and Maine).

*TOP LEFT A massive rose quartz crystal from Brazil.*

*TOP RIGHT Smoky crystal may also be known as cairngorm; this crystal was mined in Cornwall, UK.*

# Smoky Quartz *(or Smoky Topaz)*

QUARTZ GROUP

Smoky quartz, with its intricate patterns, is often cut into gemstones or *objets d'art*. Crystals weighing up to 670 lb (304 kg) have been found in hydrothermal veins in Brazil. Its distinctive smoky characteristic is probably due to rock crystal being subjected to natural radiation.

Smoky quartz is named for its smoky colour. It can be brown, black or smoky grey. Very dark crystals are called 'morion'. When heated to 572 – 752°F (300 – 400°C) crystals turn yellow then white. Quality crystals will often contain rutile inclusions.

**Range:** Smoky quartz is found worldwide. Quality crystals have been found in Brazil, Madagascar and in Alpine fissures.

---

**Mohs' Hardness** *7*
**Specific Gravity** *2.65*
**Crystal Structure Hexagonal**
**Ornamental**

---

**Mohs' Hardness** *7*
**Specific Gravity** *2.65*
**Crystal Structure Hexagonal**
**Gemstone, Ornamental**

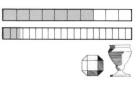

# Tiger-Eye

### QUARTZ GROUP

Tiger-eye is often used for carving-boxes and other ornamental items as these will display its distinctive markings to their full advantage. Crystals are formed from fine fibrous quartz aggregates which have had the crocidolite (a type of hornblend) altered to a yellow colour.

Tiger-eye crystals vary in colour from gold-yellow to gold-brown stripes against an almost black background. The golden hue is due to the presence of brown iron. The fibres making up the stripes are concentrated into semi-parallel groupings.

**Range:** The most important tiger-eye deposit occurs in South Africa but it is also found in western Australia, Burma, India and the United States (California).

*TOP LEFT Despite its dull and fibrous appearance, tiger-eye is often cut into slabs and polished. This example is from South Africa, the major source of tiger-eye.*

*TOP RIGHT Quartz cat's-eye is a variety of agate, which when cut appears to contain an oblique sparkle.*

# Quartz Cat's-eye

### QUARTZ GROUP

Despite its attractiveness, this material is not very valuable. It is usually cut into round polished pieces for necklaces or pendants. Quartz cat's-eye is formed from fluids associated with intrusive magmatic phenomena.

Quartz cat's-eye is semi-transparent but becomes greenish grey or green when ground. The distinctive features of the crystals are their colours and clearly fibrous appearance. Often confused with chrysoberyl cat's-eye, the material is sensitive to acids.

**Range:** Quartz cat's-eye, which usually occurs in fibrous aggregates, is found in Burma, India, Sri Lanka and West Germany.

**Mohs' Hardness** *7*
**Specific Gravity** *2.64 – 2.71*
**Crystal Structure Hexagonal**
**Gemstone, Ornamental**

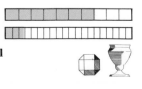

**Mohs' Hardness** *7*
**Specific Gravity** *2.65*
**Crystal Structure Hexagonal**
**Gemstone**

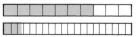

# Almandine

GARNET GROUP

Brightly coloured almandine crystals which are free of inclusions and internal cracks are sometimes cut into gemstones. Almandine is often ground and used as a medium-hard abrasive in polishing paper and cloth. Almandine crystals, which are usually well-formed, are common in medium-grade metamorphic or contact metamorphic environments.

Almandine crystals have a red colour but often contain a deep, violet-red tint. Although cut crystals have a brilliant lustre, their transparency is often marred, even in clear stones, by an excessive depth of colour. In an effort to lighten the stone the underside of such gems are often hollowed out. Unlike rubies, the deep-red colour does not lighten in natural light. The crystal splinters and fuses easily, but is insoluble in acids.

**Range:** Almandine is found in large quantities in sand deposits in Sri Lanka, with lesser deposits occurring in Afghanistan, Brazil, Czechoslovakia, Greenland, India, Madagascar, Norway, Tanzania and the United States (Alaska, California, Colorado, Connecticut, Idaho, Michigan, Pennsylvania and South Dakota).

*TOP LEFT The very deep colour of this almandine sample from the USA is typical of these crystals.*

*TOP RIGHT A fine example of grossular crystals showing growth lines and striations; from Australia.*

# Grossular

GARNET GROUP

Green and yellow tinted grossular crystals are often cut and sold as gemstones. Usually associated with metamorphism, grossular crystals have usually been weathered from their host rock and are to be found in gem gravels.

Grossular crystals are frequently green, yellow or copper-brown in colour. Cinnamon-brown and orange grossular are not uncommon. The coloration is due to iron pigments, while the green variety is caused by chrome within the crystal structure. The name grossular is derived from the Latin word *grossularia* meaning gooseberries, which the crystals sometimes resemble. The crystals vary from semi-opaque to transparent, when they display good lustre.

**Range:** Gem-quality grossular crystals are found in Canada, Kenya, South Africa, Sri Lanka (honey-yellow variety), Tanzania (green variety), Madagascar, Mexico and the United States.

**Mohs' Hardness** $6^1/_2 - 7^1/_2$
**Specific Gravity** $3.95 - 4.2$
**Crystal Structure Isometric**
**Gemstone, Industrial**

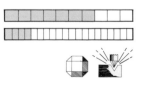

**Mohs' Hardness** $6^1/_2 - 7^1/_2$
**Specific Gravity** $3.58 - 3.69$
**Crystal Structure Isometric**
**Gemstone**

# Rhodolite

### GARNET GROUP

Rhodolite is not the most common of the red garnets, but it is the most valuable. It is formed in plutonic and ultra-mafic rocks, but due to its resistance to weathering the crystals are usually found in alluvial secondary deposits or in arenaceous rocks.

The name rhodolite is derived from the Greek word *rhodon* meaning rose coloured, and *lithos* meaning stone. The crystal itself can vary in colour from pinkish-red to rose-red and pale violet. The cut crystal displays a strong lustre and a good transparency. Rhodolite is distinguished from similar-coloured crystals of the corundum group by its lack of pleochroism or fluorescence.

**Range:** Rhodolite crystals are found in Brazil, Sri Lanka, Tanzania, Zambia, Zimbabwe and the United States (North Carolina).

*TOP LEFT This rhodolite crystal from the USA demonstrates the strong lustre of these gemstones.*

*TOP RIGHT A member of the garnet group, spessartite crystals of gem quality are rare. This sample is from Pakistan.*

# Spessartite

### GARNET GROUP

Gem-quality spessartite crystals are rare, but when they do occur, they are usually found unaccompanied by other such crystals and are well-formed. Spessartites are generally formed by, and associated with, low-grade metamorphic rocks.

Spessartite crystals are orange-pink, orange-red, red-brown or brownish yellow in colour. The crystals are unusual in that they display a high degree of hardness and density. They are semi-opaque or transparent; when transparent they have a high degree of lustre.

**Range:** The name spessartite is derived from Spessart in West Germany, where the crystals were once found. The main deposits of spessartite occur nowadays in the gem gravels of Burma and Sri Lanka. Minor deposits are also found in Brazil, Madagascar, Mexico, Sweden and the United States (California and Virginia).

**Mohs' Hardness** *6½ – 7½*

**Specific Gravity** *3.74 – 3.94*

**Crystal Structure Isometric**

**Gemstone**

**Mohs' Hardness** *6½ – 7½*

**Specific Gravity** *4.12 – 4.20*

**Crystal Structure Isometric**

**Gemstone**

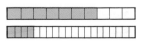

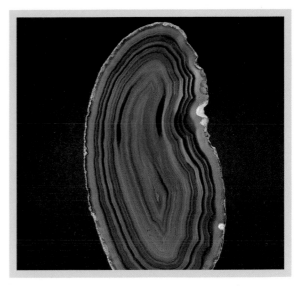

# Zircon

Zircon is an important source of zirconium, hafnium and thorium. Quality-grade crystals are cut into gem-stones with green zircons being in much demand. Zircon crystals are usually associated with intrusive acidic igneous rocks or with pegmatites derived from them. They are often to be found in alluvial deposits as pebbles or grains.

Zircon crystals – which are usually four-sided and stubby – range in colour from colourless to yellow, red, brown, grey and green. Crystals are sometimes perfectly transparent with good lustre and strong birefringence, but they can also be opaque and dull. Zircon crystals are insoluble in acids; they are infusible but very brittle.

**Range:** Gem-quality crystals are found in Cambodia, Norway, Sri Lanka, Thailand and Vietnam. Important zircon deposits also occur in Australia, Brazil, the USSR (the Urals) and the United States (Florida).

AGATE

**A**gates are believed to be a stabilizing stone, and as such can be used to calm a troubled mind or body. They are an ideal stone to use as a 'worry bead' as they have the ability to dispel anxiety. They can also be used to counter the effect of other stones which someone else may be wearing to your disadvantage.

*TOP LEFT A multicoloured zircon mass from Canada.*

*TOP RIGHT Agate from Brazil; the distinctive banding in agate is due to various impurities in a quartz-rich solution.*

# Agate

*QUARTZ GROUP*

Agate is the name given to microcrystalline quartz which is banded. These bands can be multicoloured, or as is more usual, different shades of the same colour. The former coloration gives a dramatic effect which is highly valued by collectors and jewellers.

Agates are nodules or geodes, ranging in size from a fraction of an inch to several feet in radius. They are formed from gas-created voids which have become filled with silica in a volcanic environment. The coloration is due to minute quantities of varying elements, while the banding is due to the gradual cooling of the material. Agate nodules usually have a white outer layer that is due to weathering.

Agates, and their banding, come in every conceivable colour, shade and hue. It is often a question of personal taste which agates are the more valuable; generally, those with regular widths of banding, crisp delineation between bands, and sharp vivid colours tend to appeal to both collectors and jewellers. Agates are usually cut and polished; if struck, they have a tendency to chip or splinter.

**Range:** The most important agate deposits occur in Brazil, India and Uruguay; lesser agate deposits occur in Canada, Germany, the USSR and the United States.

**Mohs' Hardness** *6½ – 7½*

**Specific Gravity** *3.9 – 4.71*

**Crystal Structure Tetragonal**

**Gemstone, Industrial**

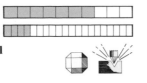

**Mohs' Hardness** *6½ – 7*

**Specific Gravity** *2.60 – 2.65*

**Crystal Structure Hexagonal**

**Gemstone, Ornamental**

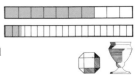

## Axinite

Axinite is named after the axe-like sharp edges which the stone usually displays when found. Quality crystals are sometimes cut into gemstones which have a modest value. Axinite is usually found within cavities in high-temperature granite or in contact metamorphic rocks.

Axinite occurs in a variety of colours: deep-brownish red, violet red, brown, rose, yellow, red-orange (manganese-rich), grey, green and occasionally blue. Crystals display a strong vitreous lustre. They have good cleavage in one direction, and poor cleavage in three directions. They are transparent to translucent and fuse easily into a glass-like, gas-filled blob.
**Range:** Axinite crystals are found in France, Japan, Mexico and the United States (Colorado, Nevada and New Jersey).

*TOP LEFT A good example of the sharp forms that axinite commonly takes; this crystal was mined in France.*

*TOP RIGHT Chalcedony in the form of botroidal overgrowths on quartz; from Cornwall, UK.*

## Chalcedony

*QUARTZ GROUP*

In ancient times chalcedony was used as a charm against idiocy and depression; it was thought to cool the blood and diffuse anger. Chalcedony, which usually forms fibrous aggregates, is the microcrystalline variety of quartz that forms concretional deposits.

The most common colours for chalcedony crystals is blue whiteish-grey. The most highly prized crystals can be brownish-yellow, red, black, green, black-and-white, grey-and-white, yellow, red, brownish red or black. Chalcedony is porous and can therefore be dyed. Natural crystals have no layering or banding.
**Range:** Chalcedony crystals are found in Brazil, India, Madagascar and Uruguay.

**Mohs' Hardness** *6½ – 7*
**Specific Gravity** *3.36 – 3.66*
**Crystal Structure Triclinic**
**Gemstone**

**Mohs' Hardness** *6½ – 7*
**Specific Gravity** *2.58 – 2.64*
**Crystal Structure Hexagonal**
**Gemstone**

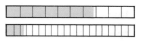

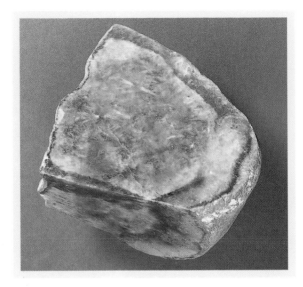

# Jade

Due to its extreme tenacity (the opposite of brittleness), jade was used in pre-historic times for making arms and tools. For this reason nephrite (a form of jade) has been called 'axe stone'. Jade has been used in China ceremonially for over 2,000 years.

Whilst jade is often considered to be a uniquely Oriental stone, it has been used in Central America for thousands of years; in pre-Columbian times it was considered to be more valuable than gold. Today, it is still used as an ornamental material, although the emergence of very high-quality synthetic jade has done a great deal to reduce the mystique which once surrounded this stone.

Jade, which is only rarely found as a crystal, is commonly sea-green, off-white or greyish-white. Occasionally, it can be brown, yellowish-brown, orange-yellow, reddish-orange, lilac, blue-grey or various combinations of grey and green. Jade is tough and translucent with a vitreous lustre. It is associated with regional metamorphosis. In its primary state jade is found in veins, but it also occurs as alluvial pebbles and large blocks.

**Range:** Jade is found in Burma, China, Guatemala, Japan, Tibet and the United States (California).

## JADE

**J**ade, as one would expect from the sea-green colour with which it is associated, is reputed to have a calming power, and as such is useful to those who suffer from any form of anxiety, tension or stress-related ailment. It is believed to establish the foundations of true friendship as well as giving one the confidence to express true love. The ancient Chinese believed that jade would guarantee a long and prosperous life.

*TOP LEFT A sectioned and polished boulder from the jade mines of Ulya River, Burma.*

*TOP RIGHT This piece of jasper, which contains traces of gold, owes its red colour to the presence of iron oxides.*

# Jasper (or Hornstone)

*QUARTZ GROUP*

The word jasper comes from the Greek word meaning 'spotted stone', and is so named because of the spotted appearance of the green variety. Jasper is often cut into slabs and used as fireplace surrounds, tables or facing material. It is a chemical, sedimentary rock which is formed as a direct result of silica precipitation, believed to have been stimulated by nearby volcanic activity. Jasper is often considered to be a form of agate.

Jasper, especially the fine grained variety, can contain up to 20 per cent of foreign material and because of this, its colour, appearance and physical properties can vary enormously, both in general terms and within single specimens. Single coloured jasper is rare: it usually occurs as irregular patterns of stripes, streaks, and spots or in a brecciated form. Jasper generally occurs as nodules and as infill in veins or cavities. Porous, it is sometimes dyed blue and sold as 'German (or Swiss) lapis lazuli'.

**Range:** Jasper is found in numerous deposits around the world. Important sources are France, Germany, India, the USSR and the United States.

---

**Mohs' Hardness** *6½ – 7*
**Specific Gravity** *3.4*
**Crystal Structure Monoclinic**
**Gemstone, Ornamental**

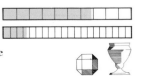

**Mohs' Hardness** *6½ – 7*
**Specific Gravity** *2.58 – 2.91*
**Crystal Structure Hexagonal**
**Ornamental**

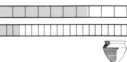

# Peridot *(or Olivine or Chrysolite)*

Good quality, clear coloured crystals are often cut and set with other gemstones. Peridot is very widely distributed in iron- and magnesium-rich igneous rocks which form a continuous series. As the amount of iron contained within the crystal structure increases its specific gravity, its solubility increases and its fusing point lowers.

Peridot crystals are typically olive green, bottle green, yellowish green or brown in colour. Crystals display a greasy vitreous lustre when split, are usually transparent with few inclusions, and are not resistant to sulphuric acid. Dark crystals can be lightened by being burned.

**Range:** Peridots are found as large crystals on the island of Zebirget in the Red Sea, in the basalt formations in the United States and in certain lavas on the Hawaiian Islands, Australia, Brazil and South Africa (where they are found alongside diamonds).

*TOP LEFT Naturally occurring gem-quality crystals of peridot from deposits in the USSR.*

*TOP RIGHT Two high-quality tanzanite specimens, showing the typical sapphire blue and amethyst colouring.*

# Tanzanite *(or Blue Zoisite)*

Good quality tanzanite crystals are much sought after by jewellers. The crystals are formed as a result of the metamorphism of plagioclases. Larger crystals are formed in high-pressure and high-temperature environments or in hydrothermal veins often associated with sulphides.

In good-quality tanzanite crystals, the colour varies from ultramarine to sapphire blue, with amethyst-colour and violet also occurring. The coloration is due to the presence of chromium and strontium within the crystal structure. When heated to 752 – 932°F (400 – 500°C), yellow and brown tints disappear and blue hues deepen. Crystals, which are usually transparent with vitreous lustre, are insoluble in acid and fuse relatively easily into a white blister-like glass.

**Range:** It is found in limited quantities in Tanzania (hence its name); these sources are now almost exhausted.

**Mohs' Hardness** *6½ – 7*
**Specific Gravity** *3.27 – 4.20*
**Crystal Structure Orthorhombic**
**Gemstone**

**Mohs' Hardness** *6½ – 7*
**Specific Gravity** *3.11 – 3.40*
**Crystal Structure Orthorhombic**
**Gemstone**

# Vesuvianite *(or Idocrase)*

Good quality green vesuvianite crystals (called cali-fornite) are popular with jewellers. They occur some-times in calcareous blocks from mafic ejections, but are normally a product of contact metamorphism.

Vesuvianite crystals occur in a variety of colours from brown to green (californite), olive-green and very occasionally yellow (xanthite), blue (cyprine), and red or white (wiluite). Crystals are usually opaque, but translucent and transparent varieties usually have a vitreous to resinous lustre. Crystals fuse easily and are virtually insoluble in acid.

**Range:** Vesuvianite was first found on Mount Vesu-vius in Italy, hence the name. Californite comes from Pakistan and the United States (California), cyprine comes from Norway, wiluite comes from the USSR, while xanthite comes from the United States (New York State).

*TOP LEFT Vesuvianite samples from Italy, where the crystal was first discovered.*

*TOP RIGHT Typically dark crystals of cassiterite on altered granite, from Cornwall, UK.*

# Cassiterite

Named after the Greek word for tin, cassiterite is an important tin ore. The crystals, which are short and prismatic, typically occur in pegmatites and greisens. The large industrial-type deposits are usually sedi-mentary in nature. They were formed in fluvial- or marine-environments, and now occur as placers. Deposits are often worked by panning, as gold often is; in this way, the heavy crystals are separated from the light material in which they are found.

Cassiterite crystals are normally brown to black in colour, but can sometimes be colourless or pink-brown. Crystals display a brilliant lustre and are in-fusible and insoluble in acids.

**Range:** The world's largest cassiterite deposits occur in Bolivia, Brazil, China, Malaysia, Sumatra and the USSR.

**Mohs' Hardness** *6½*

**Specific Gravity** *3.27 – 3.45*

**Crystal Structure Tetragonal**

**Gemstone**

**Mohs' Hardness** *6 – 7*

**Specific Gravity** *6.8 – 7.1*

**Crystal Structure Tetragonal**

**Industrial**

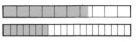

# Epidote *(or Pistacite)*

Epidote is sometimes polished and used for inlay work; other crystals are used occasionally as gemstones. Epidotes are formed in regional or contact metamorphic rocks of mafic composition.

Epidote crystals vary in colour from full-green to yellow or brown-black. The variety which is cherry-red to purplish-brown in colour contains traces of manganese, and is known as piemontite. The grey, pale-green, green-brown or pink variety contains only minute quantities of iron and is called clinozoisite. This frequently occurs as a secondary hydrothermal alteration. Epidote crystals, are transparent with a vitreous lustre and fuse fairly easily.

**Range:** Crystals are found in Austria, Norway and the United States.

*TOP LEFT Green-brown epidote crystal, frequently found as a secondary hydrothermal alteration.*

*TOP RIGHT A pale variety of hiddenite from Afghanistan; only the more distinctively coloured samples are used as gemstones.*

# Hiddenite

*SPODUMENE GROUP*

The scarcity of reasonably sized, attractive crystals makes the intensely coloured hiddenite crystal a valuable, semi-precious gem. Crystals, which are usually long and unevenly terminated, are formed in lithium-bearing pegmatites associated with quartz, feldspar, beryl and tourmaline.

Hiddenite is the green variety of spodumene, named after W. E. Hidden, the owner of the mine in the United States (North Carolina) where the crystal was first discovered in 1879. The colour of hiddenite crystals ranges from pale green to yellow-green, and from emerald-green to rich green. Hiddenite can resemble a number of other crystals (beryl, chrysoberyl, diopside, emerald or verdelite) depending on its colour. However, by testing and comparing the crystals' physical properties it should be possible to achieve a positive identification.

**Range:** The finest gem-quality crystals are found in the United States (North Carolina) while less attractive, paler crystals occur in Brazil, Burma, Madagascar and the United States (California and North Carolina).

**Mohs' Hardness** *6 – 7*

**Specific Gravity** *3.3 – 3.5*

**Crystal Structure Monoclinic**

**Gemstone, Ornamental**

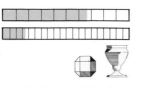

**Mohs' Hardness** *6 – 7*

**Specific Gravity** *3.16 – 3.20*

**Crystal Structure Monoclinic**

**Gemstone**

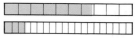

# Kunzite

SPODUMENE GROUP

Kunzite crystals with few inclusions and good transparency are often cut into gemstones. Kunzite, which is an important source of lithium and its salts, is formed in lithium-bearing pegmatites associated with quartz, feldspar, beryl and tourmaline.

Kunzite is the pink-violet, light-violet, green-violet or brown variety of spodumene. The crystal is named in honour of George F. Kunz (1850–1932), the mineral collector who first described it at the turn of this century. The kunzite crystals, which are usually long and unevenly terminated, generally have few inclusions. They are transparent, and display a marked pleochroism which is seen as a difference in colour depth in different directions.

**Range:** Kunzite crystals are found in Brazil, Madagascar and the United States (California, Connecticut, and Maine).

KUNZITE

**K**unzite crystals are considered essentially feminine in shape and purpose. They help rejuvenate the skin and are good for the heart. They can stimulate self-praise and fulfilment and are reputed to regulate the menstrual cycle.

*TOP LEFT This rose-coloured kunzite crystal is from Afghanistan. The striations are due to post-formation earth movements and are not growth related.*

*TOP RIGHT A typically squat formation of amazonite, one of the feldspar group.*

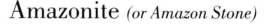

# Amazonite (or Amazon Stone)

FELDSPAR GROUP

Amazonite, which is frequently confused with jade or turquoise, is often ground into beads for necklaces or fashioned into ornamental objects. Amazonite crystals, which are normally squat and slightly prismatic, are found in metamorphic, intrusive and pegmatic rocks.

Deriving its name from the Amazon River, and the belief that the crystals somehow flowed from it, the amazonite crystal is usually light green but sometimes blue-green or bluish. Amazonite usually has a mottled appearance and sometimes has a fine criss-cross network of light striations. Crystals are generally semi-opaque, with poor lustre and easy cleavage.

**Range:** Important amazonite deposits occur in Australia, Brazil, India, Madagascar, Namibia, the USSR, the United States and Zimbabwe.

**Mohs' Hardness** *6 – 7*
**Specific Gravity** *3.16 – 3.20*
**Crystal Structure Monoclinic**
**Gemstone, Industrial**

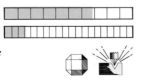

**Mohs' Hardness** *6 – 6½*
**Specific Gravity** *2.56 – 2.58*
**Crystal Structure Triclinic**
**Gemstone, Ornamental**

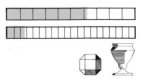

# Benitoite

Benitoite crystals, which are sometimes confused with light coloured sapphires, are often cut into gemstones. The crystals are found in veins in the brecciated (or fragmented) body of a blue schist associated with serpentinite.

Benitoite crystals, which are usually stubby and zoned, can range in colour from light to dark blue. Crystals look blue when viewed through the acute faces of the rhombohedron and colourless when viewed through the obtuse faces.

**Range:** Benitoite is named after the only deposit found to date, which occurs in San Benito county, California. Modern exploration techniques, make it likely that it will not be long before other deposits are discovered.

*TOP LEFT Benitoite from California, USA, the only location where this crystal has yet been found.*

*TOP RIGHT The play of colour in this labradorite sample is due to the differential interference of light on twinned layers; this specimen is from Norway.*

# Labradorite (or Spectrolite)

FELDSPAR GROUP

Labradorite is usually fashioned into decorative boxes or *objets d'art*, where its colours can be displayed to their full advantage. Smaller specimens are sometimes made into beads, brooches or ring stones. Labradorite is typically associated with eruptive and metamorphic rocks.

Labradorite may at first appear to be a dark smoke-grey colour, but when the light strikes it in a certain way, it displays rainbow-like colours (violet, indigo, blue, green, yellow, orange and red) in an effect similar to that of gasoline lying on water. The most distinctive feature of labradorite is its iridescence against a dark background. This effect is probably caused by the interference of light on twinned lamellae (or layers).

**Range:** Labradorite usually occurs as a compact aggregate and rarely as an individually formed crystal. The world's most spectacular specimens come from Finland, with lesser rocks coming from Canada (Labrador, hence the name) Madagascar, Mexico, the USSR and the United States.

**Mohs' Hardness** *6 – 6½*

**Specific Gravity** *3.65 – 3.68*

**Crystal Structure Hexagonal**

**Gemstone**

**Mohs' Hardness** *6 – 6½*

**Specific Gravity** *2.62 – 2.76*

**Crystal Structure Triclinic**

**Gemstone, Ornamental**

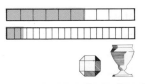

# Moonstone
## (or Adularia Moonstone)

*FELDSPAR GROUP*

Moonstones, which display a blue reflection, are highly prized by jewellers. They could be confused however, with heat-treated amethysts or with milky synthetic spinel, if it were not for the fact that these two crystals do not display the correct mobile reflective characteristics. Moonstone is a variety of feldspar formed in association with orthoclase and albite (with a predominance of orthoclase).

In general, moonstones are almost colourless, having only a pale-grey or yellow tint, with a whitish to silvery-white or blue shimmer. Incipient cleavage cracks are sometimes visible within the crystals; they have a slight turbidity and a distinctive mobile reflection.

**Range:** Important deposits of moonstone occur in Australia, Burma, India, Sri Lanka, Tanzania and the United States.

---

MOONSTONE

**T**he misty silver colour of the moonstone is said to brighten at the start of every new moon. When passed between lovers, the moonstone is reported to arouse a feeling of warmth and friendship which many physical relationships fail to develop. The moonstone is a sensitive stone and is very susceptible to mood changes.

*TOP LEFT A sample of cut moonstone crystal of the sort valued highly by jewellers.*

*TOP RIGHT Prehnite crystals demonstrate a vitreous lustre; this example is from Scotland.*

---

# Prehnite

Prehnite is of interest to geologists as an indicator of formation sequences. Impressive crystals of prehnite can be found in museums and in mineral collections. Prehnite crystals have usually crystallized out from hydrothermal fluids in cavities contained within basaltic volcanic rocks.

Prehnite crystals are rare but when they occur they are usually white-green, light-green, green-yellow or yellow-brown. They can also occur as similar coloured stalactitic aggregates. The crystals display perfect basal cleavage and are transparent with vitreous lustre. They fuse easily, and dissolve slowly in hydrochloric acid with no gelatinous-silica residue.

**Range:** Crystals have been found in Australia (New South Wales), China, France, South Africa and the United States (Lake Superior, New Jersey and Virginia).

---

**Mohs' Hardness** *6 – 6½*
**Specific Gravity** *2.56 – 2.62*
**Crystal Structure Monoclinic**
**Gemstone**

**Mohs' Hardness** *6 – 6½*
**Specific Gravity** *2.87 – 2.95*
**Crystal Structure Orthorhombic**
**Scientific**

# Pyrite

The Incas in Central America used slabs of polished pyrite as mirrors. Nowadays it is used in the manufacture of sulphuric acid by the lead chamber method, which is used frequently for producing fertilizers. Pyrite commonly occurs in plutonic, volcanic, sedimentary and metamorphic rocks. Pyrite crystals also occur in hydrothermal and medium- to low-temperature quartz veins.

Pyrite gets its name from the Greek word for fire, as it sparks when struck. Pyrite crystals tend to be brass-yellow or grey-yellow with a strong metallic lustre. Due to its appearance, it is often referred to as 'fool's gold'. Pyrite is insoluble in hydrochloric acid; when powdered it will dissolve in nitric acid.

**Range:** Pyrite is found throughout the world. Large deposits are found in Germany, Italy, Japan, Norway, Spain, Sweden and the United States (Arizona, Colorado, Illinois and Pennsylvania).

*TOP LEFT Striated pyrite crystals from the USA, showing the metallic lustre typical of these crystals.*

*TOP RIGHT Orthoclase sample from Cornwall, UK; this crystal is an important source of porcelain.*

# Orthoclase

*FELDSPAR GROUP*

Orthoclase is an important industrial mineral. When mixed with kaoline and quartz, it can be formed or moulded easily, and then fired at low temperatures into porcelain. In its purest state, orthoclase can be made into porcelains suitable for high-tension electrical insulators, ceramic glazes and dental products. Orthoclase sometimes forms crystals, known as noble orthoclase, which are cut and polished for use as semi-precious gemstones. Orthoclase commonly occurs in intrusive, magmatic, and metamorphic rocks, where it has cooled slowly.

Orthoclase can be transparent or yellow (as in the case of noble orthoclase crystals), but is usually semi-opaque or white to greyish-white, yellowish-white or reddish. The mineral is so called because it has two directions of cleavage which are orthogonal (at right angles) to each other. Crystals are coloured a flame violet when potassium is present.

**Range:** Orthoclase is widely distributed. Important sources occur in Czechoslovakia, Italy, Spain, Switzerland and in various locations in the United States. The best examples of noble orthoclase come from Madagascar.

**Mohs' Hardness** $6 - 6\frac{1}{2}$
**Specific Gravity** $5 - 5\frac{1}{2}$
**Crystal Structure Isometric**
**Industrial**

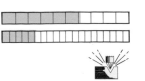

**Mohs' Hardness** $6$
**Specific Gravity** $2.55 - 2.63$
**Crystal System Monoclinic**
**Industrial, Gemstone**

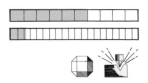

# Hematite
## *(or Bloodstone or Specularite)*

Hematite used to be fashionable as mourning jewellery, but it is now used as a ring stone and in bead necklaces. In ancient times, hematite was used as a talisman against bleeding. It is the most important source of iron-ore, and is used also as a pigment and a polishing powder. Hematite is a fairly common mineral which is deposited in hydrothermal veins or in lavas, where it forms under oxidizing conditions.

Hematite crystals are blood-red in thin sections, but are normally black or dark iron-grey with a strong metallic lustre. Hematite is a heavy crystal; although fragile, it does not have any cleavage planes. The crystals dissolve slowly when heated in concentrated hydrochloric acid. They are infusible and become magnetic when heated in a reducing flame.

**Range:** The world's largest hematite deposits occur in Angola, Brazil, Canada (Quebec) and the United States (Lake Superior). Iron-rose crystals are found in St Gotthard in Switzerland and Minas Gerais in Brazil where specimens up to 6 inches (15 cm) in diameter have been found.

---

### HEMATITE

**H**ematite (or bloodstone), as may be expected, is associated with blood disorders and purification. It is reported to stimulate courageous behaviour, both physical and emotional, and to restore the balance to highly charged situations. Women may find that the bloodstone is useful for regulating irregular menstrual flow.

---

*TOP LEFT Hematite is a common source of iron-ore; this piece is from Cumbria, UK.*

*TOP RIGHT As its name implies, magnetite is the most magnetic mineral known. This sample was mined in Italy.*

# Magnetite *(or Lodestone)*

Magnetite was used by the Vikings as a navigational aid: by placing a length of magnetite on a piece of wood floating in a bowl of water, they created the first compass. They used this crude but effective device to navigate their longboats from Scandinavia (via Iceland and Greenland) to Nova Scotia and Newfoundland. Magnetite has the highest iron content of any of the iron ores. Vanadium and phosphorous are often collected from the slag when smelting magnetite. Magnetite is a common constituent of most igneous rocks; it sometimes occurs as black beach sands.

Magnetite is iron-black or steely-grey in colour and usually occurs as a compact and granular mass. It is strongly magnetic (hence the name) and infusible, but will dissolve slowly in concentrated hydrochloric acid.

**Range:** The largest magnetite deposits occur in Kiruna in northern Sweden where the same ore deposit has been mined for over a century. Secondary deposits occur in the USSR and the United States (New York State and Utah).

---

**Mohs' Hardness** *5½ – 6½*
**Specific Gravity** *4.95 – 5.30*
**Crystal Structure Hexagonal**
**Gemstone, Industrial**

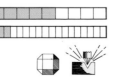

**Mohs' Hardness** *5½ – 6½*
**Specific Gravity** *5.2*
**Crystal Structure Isometric**
**Industrial**

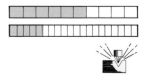

## Opal *(or Precious Opal)*

*QUARTZ GROUP*

Opal is used as a gemstone because it contrasts well with other items of jewellery. It is never found in visible crystals, although recent work using electron microscopes has shown that the opal consists of minute balls of the mineral cristobalite suspended in a jelly-like matrix of silica. Opals occur as fine-grained masses occupying voids and veins, usually in a grape-like shape.

Opals can be divided into two groups: white or light coloured opals, which are known as white or milky opals; and the rarer black opals which can range in colour from dark green to dark grey, and from dark blue to grey-black or totally black. Opals are extremely porous, and since they can contain up to 30 per cent of water by volume, they age quickly by giving up this water. They fracture and chip easily, and are super-sensitive to acids, alkalis and heat.

**Range:** Opals have been mined in Czechoslovakia since Roman times, and, until the nineteenth century this was the only source of the stone. Today most of the high quality opals come from the Lightning Ridge area of New South Wales (Australia). Opals also occur in Guatemala, Honduras, Mexico and the United States (California, Idaho, Nevada, Oregon and Wyoming).

OPAL

Due to the ease with which opals will give up or take up water, they are often considered to be a living stone which can move forward in the future. It is not surprising that some people believe that such journeys age the opal beyond its limitations and causes ageing, which leads to it cracking. Opals are a feminine stone, once thought to guard against grey hair in blonde women. It was also believed to make childbirth less painful. Due to their cantankerous nature, opals should be kept away from other stones.

*TOP LEFT This unusual example of a fire (precious) opal is from Mexico. Never found as a visible crystal, opals have a high water content.*

## Rhodonite

Rhodonite is often used to make ornamental figures and large *objets d'art*; these are particularly valuable when veined by black manganese oxide. Rhodonite is formed through the metamorphic action of impure limestones which are themselves the result of element substitution involving magmatic fluids.

Rhodonite crystals tend to be pink, reddish-pink, or brown with veins or patches of black caused by the presence of manganese oxides. Rhodonite crystals display perfect prismatic cleavage (almost at right angles), and are transparent to translucent with a vitreous lustre. Rhodonite is insoluble in acid, unlike the similar but softer rhodochrosite.

**Range:** Rhodonite is fairly widespread; quality crystals are found in Australia, Brazil, Finland, India, Japan, Madagascar, Mexico, New Zealand, South Africa, the USSR (the Urals) and the United States.

*TOP RIGHT Rhodonite of a characteristic pink tone.*

**Mohs' Hardness** *5½ – 6½*
**Specific Gravity** *1.98 – 2.20*
**Crystal Structure Non-crystalline**
**Gemstone**

**Mohs' Hardness** *5½ – 6½*
**Specific Gravity** *3.4 – 3.7*
**Crystal Structure Triclinic**
**Ornamental**

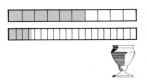

# Sodalite

Sodalite is used in bead necklaces and *objets d'art*. It can easily be confused with lapis lazulite, especially as pyrite occurs in both. Sodalite occurs in under-saturated plutonic rocks, associated with meta-morphosed limestones and volcanic blocks.

The colour of sodalite varies from bright blue to violet, and from white to grey with green tints. The white is usually due to calcite. Sodalite crystals are fragile with poor cleavage, and translucent with vit-reous lustre. They are soluble in hydrochloric and nitric acid and leave a silica gel when dissolved. Be-cause of the presence of sodium, the crystal fuses easily giving a yellow flame.

**Range:** The most important sodalite deposits are in Brazil (Bahia) and Canada (Ontario). Less substantial deposits occur in Bolivia, Burma, Greenland, India, Portugal, Romania, the USSR and Zimbabwe.

### LAPIS LAZULI

**L**apis Lazuli is often con-sidered a suitable stone for starting one's spiritual birth. It is also believed to act as a medium through which positive psychic influences are transmitted. With a history of use going back over 7,000 years, its powers to develop inner strength and con-fidence are legendary.

*TOP LEFT Sodalite is used to make jewellery and ornaments.*

*TOP RIGHT Pure specimens of lapis lazuli such as this are prized by jewellers and artisans.*

# Lapis Lazuli
### (or Lapis or Lazulite)

Lapis Lazuli has been used as an ornamental stone for thousands of years; it is often fashioned into large *objets d'art* or else cut into slabs for facing purposes.

During the Middle Ages, lapis lazuli was often powdered and used as a pigment; today this pigment (produced from synthetic stone) is used in ultramarine paint. Lapis lazuli is an uncommon mineral, which usually occurs as compact masses associated with metamorphosed limestone (marble).

The name lapis lazuli is a hybrid term deriving from both Latin and Arabic meaning 'blue-stone' (the stone has a distinctive azure hue). Lapis lazuli is a complex mineral and consists of a combination of several minerals, including calcite, augite, mica, pyrite and pyroxene. Although it polishes well, it is fragile and will break imperfectly. It is porous and may be treated with paraffin to enhance the blue colour.

**Range:** The best quality lapis lazuli comes from Afghanistan. Chile and Russia also produce large quantities of the stone, while small deposits also occur in Angola, Burma, Canada (Labrador), Pakistan and the United States (California and Colorado).

**Mohs' Hardness** *5½ – 6*
**Specific Gravity** *2.13 – 2.33*
**Crystal Structure Isometric**
**Gemstone, Ornamental**

**Mohs' Hardness** *5 – 6*
**Specific Gravity** *2.38 – 2.9*
**Crystal Structure Isometric**
**Ornamental**

# Brazilianite

Brazilianite is a rare and beautiful crystal which is much prized by jewellers and collectors. The crystals occur in cavities in pegmatites associated with blue apatite, clay and lazulite.

Brazilianite crystals, which are usually elongated and stubby prisms, are yellow or green-yellow. They are often quite large. The crystals are fragile with perfect cleavage, transparent with a vitreous lustre. They will dissolve with difficulty in strong acids, and will only fuse in small fragments. They quickly lose their colour when heated.

**Range:** As the name suggests, Brazil is the major, and was until relatively recently, the only source of this crystal. It has now been found in the United States (New Hampshire) as well.

*TOP LEFT A sample of brazilianite from Brazil, until recently the only known source of this rare crystal.*

*TOP RIGHT Chromite from Iran: this crystal has a variety of industrial applications.*

# Chromite

Chromite is the most important ore for chrome. Due to its hardness and resistance to chemical attack, it is used for plating and alloying with other metals. This produces hard chrome steel, stainless steel and nickel-chrome. Chromium is used in the dying and tanning of leather, for producing yellow pigment and plays a significant part in the manufacture of refractory bricks. It occurs as a primary mineral of ultrabasic rocks and sometimes in deposits which have been transported and later deposited in a different location.

Chromite rarely occurs as individually formed crystals, but more commonly as a brown-black or black submetallic granular mass. It is weakly magnetic, infusible and insoluble in acids.

**Range:** Chromite occurs in quantities worthy of economic extraction in Cuba, India, New Caledonia, South Africa, Turkey and the USSR (the Urals).

**Mohs' Hardness** *5½*
**Specific Gravity** *2.98 – 2.99*
**Crystal Structure Monoclinic**
**Gemstone**

**Mohs' Hardness** *5½*
**Specific Gravity** *4.5 – 4.8*
**Crystal Structure Isometric**
**Industrial**

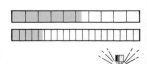

# Enstatite

## PYROXENE GROUP

Enstatite is of interest to geologists, scientists and crystal collectors, as its resistance to melting means it is often a good indicator of the direction of original lava flow. It is formed in mafic and ultra-mafic, plutonic and extrusive volcanic environments, as well as in high-grade metamorphic rocks.

Enstatite crystals, are sometimes stubby and prismatic; more often, they are fibrous or plate-like masses. They come in a variety of colours, ranging across yellowish, green, olive-green, grey-green or grey. The green-brown variety has a high iron content and is called bronzite. The name enstatite comes from the Greek word meaning very resistant to melting, and as one would expect it is insoluble in acids and almost infusible. Crystals display good cleavage; they are transparent with a vitreous lustre changing to pearly on cleavage surfaces.

**Range:** Enstatite is found in Germany, India, Japan, Northern Ireland, Norway, Scotland, South Africa, the USSR and the United States.

*TOP LEFT This enstatite sample from Norway demonstrates the vitreous lustre and grey-green colouring that is typical of the crystal.*

*TOP RIGHT Nickeline or niccolite is found in high-temperature hydrothermal veins; this piece is from South Africa.*

# Nickeline *(or Niccolite)*

Nickeline was the first mineral from which nickel was derived. It is still employed for this purpose when sufficiently large masses occur. Nickel is of major importance as an alloying metal. It is used in German silver (nickel, copper and zinc), nichrome (nickel, chrome and iron), nickel steel (nickel and iron), money metal (25 per cent nickel and 75 per cent copper) and stainless steel. Nickel alloys possess a number of useful characteristics: they are ductile, strong, resistant to chemical attack and heat stable. Nickeline occurs in high-temperature hydro-thermal veins.

Nickeline crystals are rare, but, when they do occur, they are squat, plate-like or pyramidal. More frequently, they occur as a compact metallic bronze mass with pink, blue, grey and sometimes turquoise colorations. It fuses easily giving off a strong, garlic-like smell. Nickeline is soluble in nitric acid producing a green solution.

**Range:** Crystals have been found in Germany, while large masses of nickeline occur in Argentina, Canada, Germany and Japan.

**Mohs' Hardness** *5½*
**Specific Gravity** *3.26 – 3.28*
**Crystal Structure Orthorhombic**
**Scientific**

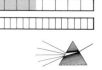

**Mohs' Hardness** *5 – 5½*
**Specific Gravity** *7.5 – 7.8*
**Crystal Structure Hexagonal**
**Industrial**

## Diopside

PYROXENE GROUP

Despite the fact that diopsides are often confused with emeralds, hiddenites or peridotes, they are popular with jewellers; transparent specimens often are cut into gems. Diopside crystals are formed from contact metamorphism, especially in dolomitic marbles which are associated with other calcium silicates.

Diopside crystals are usually white, yellow, green, blue, brown or, very occasionally, colourless. The purple manganese-bearing variety is known as violane; the dark-green chromium-bearing variety is known as chromian diopside; and the vanadium-bearing variety is known as lavrovite. Crystals are fragile and have perfect prismatic cleavage. They are transparent to translucent with a vitreous lustre. They tend to fuse easily but are insoluble in acids.

**Range:** Diopside crystals are found in Austria, Finland, Greenland, Madagascar, South Africa and Sweden.

*TOP LEFT A yellow form of diopside popular with jewellers; this one was found in Quebec, Canada.*

*TOP RIGHT An attractive blue form of lazulite from the USA.*

## Lazulite

Lazulite is a gemstone of little importance. It is found in silica-rich rocks, pegmatites and in quartz veins associated with andalusite and rutile or metamorphic rocks.

Lazulite crystals, which are rare, can be dark blue, bright blue or blue-white. They possess a distinctive prismatic cleavage and are transparent with a vitreous lustre. They will dissolve with difficulty in hot concentrated acid, are infusible, but will discolour and crumble when heated.

**Range:** Lazulite is found in Austria, Brazil, India, Madagascar, Sweden, Switzerland and the United States (California, Georgia, Maine and North Carolina).

**Mohs' Hardness** *5 – 6*
**Specific Gravity** *3.27 – 3.31*
**Crystal Structure Monoclinic**
**Gemstone**

**Mohs' Hardness** *5 – 6*
**Specific Gravity** *1.61 – 1.65*
**Crystal Structure Monoclinic**
**Gemstone**

# Turquoise

In ancient times turquoise was used in jewellery and in the preparation of cosmetics. The stone is now used primarily for making necklaces, brooches, amulets and *objets d'art*. Turquoise is a secondary mineralization, due to alteration in arid regions of aluminium-bearing rocks. It is rich in apatite and chalcopyrite.

Turquoise comes in a range of colours from blue-white to sky-blue, and light greenish blue to light blue; it is generally opaque. Surprisingly, turquoise was not thought to be crystalline until crystals were found in Virginia, in the United States, in 1911. Sky-blue crystals turn dull green when heated at 482°F (250°C). Turquoises, which usually occur as grape- or kidney-shaped aggregates, are soluble in hydrochloric acid.

**Range:** Turquoise means 'Turkish-stone', and is so called because the stones used to be brought to Europe along the overland trading route through Turkey. The best quality turquoise comes from Iran, with prolific deposits also being worked in the United States (Arizona, California, Nevada and New Mexico).

## TURQUOISE

**T**urquoise is considered, by some North American Indians, to be a constant reminder that man is merely a spirit contained within a human form. The stone is associated with courage and strength. It is sacred to the Pueblo Indians of New Mexico, and, until the end of the 17th century, it could only be worn by men of that tribe. Because of its high copper content, it is a wonderful conductor of healing energies.

*TOP LEFT Turquoise has been used to make jewellery for centuries; these turquoise nuggets are from Arizona, USA.*

*TOP RIGHT Uraninite is an important source of uranium, and was the original source of radium. This piece was mined in Cornwall, UK.*

# Uraninite *(or Pitchblende)*

Uraninite is the main source for uranium and radium. The Curies (Marie (1867–1934), and her husband, Pierre (1859–1906) first identified polonium, radium and helium, from a sample of pitchblende. These elements were known even then to occur in the sun, but had still not been isolated and identified back on earth. Uranium is used in power generation, in nuclear fusion, and as a source of radium whose radioactive qualities are used in many branches of security. It has many uses in medicine, and in monitoring (fluid density, speed, quantity, etc). Uraninites often occur in pegmatites, medium- or high-temperature hydro-thermal veins, or concentrated in gold-bearing conglomerates due to their high specific gravity.

Uraninite occasionally occurs as a dull-black cubic crystal, but more frequently is found as a granular mass or aggregate which commonly displays a bright yellow or orange colour, or combinations of the two. It is a very heavy mineral, fragile but with no predictable cleavage patterns; is highly radioactive, infusible, and will dissolve in most acids except hydrochloric acid.

**Range:** Uraninite is mined in Australia, Canada, France, Namibia, South Africa and the United States (Arizona, Colorado and Utah).

**Mohs' Hardness** *5 – 6*
**Specific Gravity** *2.6 – 2.9*
**Crystal Structure Triclinic**
**Gemstone, Ornamental**

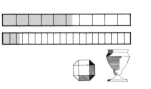

**Mohs' Hardness** *5 – 6*
**Specific Gravity** *7.5 – 10*
**Crystal Structure Isometric**
**Industrial, Scientific**

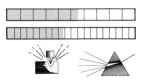

## Apatite

Apatite is used in the manufacture of phosphate fertilizers and in the chemical industry to make salts of phosphoric acid and phosphorous. It is associated with all types of eruptive rocks, in some hydrothermal veins and in iron-rich igneous rocks. It is also common in marine sedimentary rocks where it has been formed by chemical deposition.

Apatite crystals vary considerably in colour from colourless to yellow, from green to brown and occasionally from red to violet or blue. The crystals are very fragile with poor cleavage parallel to the base. They tend to be transparent to opaque with a vitreous lustre. Some crystals lose their colour when heated while others will fluoresce a bright yellow in ultraviolet light.

**Range:** Apatite is a common material and is found in Austria, Canada (Ontario), Chile, Mexico, Morocco, South Africa, Togo, Tunisia, Naura and the United States (Florida and Maine).

*TOP LEFT Both translucent and opaque crystals of apatite can be seen in this sample from Portugal.*

*TOP RIGHT Dioptase crystals from the oxidized part of a copper bearing lode in Tsumeb, southern Africa.*

## Dioptase

Dioptase is used for making jewellery and is popular with crystal collectors. It occurs as short prismatic crystals within cavities in the oxidation zones of copper deposits.

Dioptase crystals are a bright emerald-green colour. They are fragile with perfect cleavage, transparent to translucent with a vitreous lustre. Although infusible, will expand and turn black when heated. Dioptase crystals are soluble in ammonia; they dissolve in hydrochloric acid and nitric acid, leaving a silica residue.

**Range:** Dioptase is found in Chile, Namibia, the USSR, the United States (Arizona) and Zaire.

**Mohs' Hardness** 5
**Specific Gravity** 3.16 – 3.23
**Crystal Structure Hexagonal**
**Industrial**
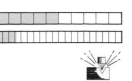

**Mohs' Hardness** 5
**Specific Gravity** 3.28 – 3.35
**Crystal Structure Hexagonal**
**Gemstone**
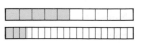

# Hemimorphite

Hemimorphite is primarily a zinc ore, but fine specimens can either be cut and set as gemstones or used for decorative fabrication in much the same way as the similar-looking turquoise. Hemimorphite is formed in the oxidized zones of lead and zinc deposits.

Hemimorphite is rarely found as large crystals, but more commonly as platex crystals whose ends are different or hemimophic (hence the name). Crystals, when they do occur, tend to be white, transparent or translucent, but are more often tinted a yellow, green, blue or brownish hue by the presence of copper or iron. In North America, hemimorphite is often referred to as calamine while in Europe smithsonite is often called calamine. It is soluble in strong acids but fuses with difficulty.

**Range:** Hemimorphite deposits are found in Algeria, Greece, Italy, Mexico, Namibia and the United States (Colorado, Montana and New Jersey).

*TOP LEFT Pale blue hemimorphite crystals, their colour imparted by the presence of copper; from Mexico.*

*TOP RIGHT A good sample of translucent scheelite from Cumbria, UK.*

# Scheelite

Scheelite is an important source for tungsten, which is used for alloying in order to produce a very strong steel. Scheelite crystals are formed in high-temperature pegmatitic and hydrothermal veins.

Scheelite crystals range in colour from no colour at all to yellow (due to molybdenum traces), and from orange to brown. The crystals, which tend to be striated, are fragile with good cleavage and tend to be translucent or transparent with vitreous lustre. Scheelite is soluble in acids and fuses with difficulty.

**Range:** Where industrial exploitation is concerned, the most important deposits occur in Australia, Bolivia, Burma, China, Japan, Malaysia and the United States.

**Mohs' Hardness** *4¹/₂ – 5*
**Specific Gravity** *3.4 – 3.5*
**Crystal Structure Orthorhombic**
**Gemstone, Ornamental**

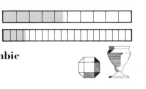

**Mohs' Hardness** *4¹/₂ – 5*
**Specific Gravity** *5.9 – 6.1*
**Crystal Structure Tetragonal**
**Industrial**

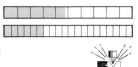

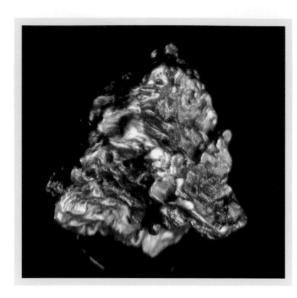

# Platinum

When the electronics industry was expanding in recent decades, platinum's ability to conduct electricity without ozidizing lead to a renewed interest in this neglected metal. Interest grew when it was realized that there were many uses to which platinum (and its associated platinum group metals – or PGMs) lent itself as a chemical catalyst. Today, the world's need for PGMs as catalytic exhausts has contributed greatly to the unprecedented demand for platinum.

Platinum is formed by the action of hydrothermal fluid intrusion in basic and ultra-basic igneous rocks; it is also found in depositional areas derived from these rocks. The crystal usually occurs in rounded grains or scales, and very occasionally as nuggets. It is malleable with a metallic silver-grey lustre. Platinum is very heavy and may be weakly magnetic. It will not dissolve in acids (except aqua regis) and only fuses at 3,227°F (1,775°C.)

**Range:** The world's major source of platinum is the Witwatersrand region of the Transvaal (South Africa). Smaller quantities of platinum are found in Australia, Canada (Ontario), the USSR (the Urals) and the United States (Alaska).

*TOP LEFT A rare nugget of platinum, taken from the Yukon, USA.*

*TOP RIGHT Kyanite is one of the crystals which has a different hardness along its length to that across its width.*

# Kyanite *(or Disthene)*

Kyanite is a major raw material required for the manufacture of high-temperature porcelain products, perfect electrical insulators and acid-resistant products. Formed in perlitic rocks rich in aluminium and metamorphosed under high pressure, kyanite crystals are usually long, flat and prismatic.

Kyanite crystals vary in colour from colourless to grey, and from green to blue-green or blue. Crystals are fragile, display perfect cleavage and are transparent or translucent, with a pearly lustre on cleavage planes. Crystals are infusible and are insoluble in acid.

**Range:** Kyanite is found in Australia, Austria, France, India, Kenya, Switzerland and the United States (Connecticut, Massachusetts and North Carolina).

**Mohs' Hardness** *4 – 4½*
**Specific Gravity** *14.0 – 19.0*
**Crystal Structure Isometric**
**Industrial**
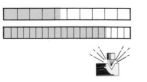

**Mohs' Hardness** *4 – 5 along axis, 6 – 7 across axis*
**Specific Gravity** *3.65 – 3.69*
**Crystal Structure Triclinic**
**Industrial**
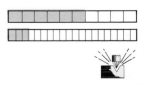

# Smithsonite
## *(or Bonamite or Dry-bone Ore)*

*TOP LEFT Smithsonite rarely occurs in crystal form; this unusual example is from Tsumeb, southern Africa.*

*TOP RIGHT Nodules of variscite taken from mines in Devon, UK.*

Smithsonite was named after James Smithson (!765–1829), the founder of the world-famous Smithsonian Institute in Washington DC. It is used as a source of zinc and samples which have particularly striking bands will often be cut, polished and used for ornamental purposes. Smithsonite occurs in voids and as stalactite-like layers in limestone cavities.

Smithsonite is white when pure, but will adopt any number of colours depending on the presence of additional elements or minerals: copper (usually malachite) will give a green to blue-green colour; cadmium will give a bright yellow colour; and cobalt a pink colour. Combinations of the above with the addition of small quantities of iron, manganese, magnesium or lead can give brown or violet crystals.

Smithsonite sometimes occurs as crystals, but is often found in solid (sponge-like) form resembling dry bones – hence the name bonamite or dry-bone ore. Smithsonite is infusible, but soluble (with effervescence) in cold hydrochloric acid.

**Range:** Smithsonite deposits occur in Australia, Austria, England, Greece, Namibia, Sardinia, Spain, Turkey, the USSR and the United States (Arkansas, Colorado and New Mexico).

# Variscite *(Utahlite)*

Variscite, which is often confused with turquoise, is used as an ornamental stone. It is formed by the infusion of phosphatic waters into aluminous-rich rocks. Variscite rarely occurs as crystals but more normally as nodules and masses.

Variscite is usually pale-green, yellow-green or green with a blue tint. Crystals tend to display no cleavage but break easily, giving concoidal fractures and very smooth surfaces. It is translucent with a vitreous to waxy lustre, and the crystals are infusible but will discolour if heated.

**Range:** Nodules over 3½ feet (107 cm) in diameter have been found in Utah (hence the name). Quality stones are also found as nodules and masses in Austria, Bolivia and the United States (Arkansas and Nevada).

**Mohs' Hardness** 4 – 5
**Specific Gravity** 4.3 – 4.5
**Crystal Structure Hexagonal**
**Ornamental, Industrial**

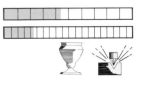

**Mohs' Hardness** 4 – 5
**Specific Gravity** 2.4 – 2.6
**Crystal Structure Orthorhombic**
**Ornamental**

# Fluorite (or Fluorspar)

### HALIDE GROUP

Fluorite is one of the mainstays of the modern chemical industry. It is used in the production of hydrofluoric acid, which is essential for petrochemical production, in plastics manufacturing and oil-well stimulation.

Fluorite is usually associated with medium- and high-temperature hydrothermal veins, while the best quality crystals have generally grown in cavities or been deposited in basins containing salt-rich waters.

The colour of fluorite crystals varies enormously: from colourless and completely transparent when pure, to yellow, green, pink, purple or violet, and blue or black when certain elements become tied up within the crystal structure. Fluorite crystals will fluoresce (hence the name fluorspar) a bright violet or blue colour. The crystals, which usually have an uneven or splotchy distribution of hues, are insoluble in water and most acids, with the exception of concentrated sulphuric acid. Crystal edges will fuse fairly easily, turning the flame brick red.

**Range:** Colourless crystals are found in Italy, pink crystals in Switzerland, green crystals in Norway, and violet crystals in England. Mineable reserves of fluorite are found in Canada, England, Italy, Mexico, the USSR and the United States (Illinois and Kentucky).

*TOP LEFT The pure form of fluorite is colourless, but samples such as this from the UK owe their colour to traces of impurities.*

*TOP RIGHT A pyrrhotite mass, showing layered growth; from Mexico.*

# Pyrrhotite

Pyrrhotite, by itself, is only of use and interest to mineral collectors, but is more often than not associated with copper, iron and sulphur. It is also a major source of cobalt, nickel and platinum. Pyrrhotite is fairly common in mafic and ultramafic extrusive rocks, but also occurs in pegmatites, high-temperature hydrothermal veins, and occasionally as a sedimentary deposit.

Pyrrhotite crystals are usually tabular with striated faces. It fuses easily, changing from slightly magnetic to strongly magnetic, and gives off hydrogen sulphide fumes when dissolved in hydrochloric acid. Pyrrhotite frequently occurs as a granular mass which is a reddish-brown colour with a shiny metallic bronze lustre.

**Range:** Substantial pyrrhotite deposits occur in Canada (Manitoba and Ontario). Quality crystals have been found in Brazil (Minas Gerais), Mexico and the United States (Maine and New York State).

**Mohs' Hardness** *4*

**Specific Gravity** *3.1 – 3.33*

**Crystal Structure Isometric**

**Industrial**

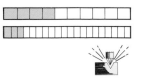

**Mohs' Hardness** *3½ – 4½*

**Specific Gravity** *4.6 – 4.7*

**Crystal Structure Hexagonal**

**Scientific, Industrial**

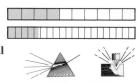

# Rhodochrosite *(or Inca-Rose)*

## CALCITE GROUP

Rhodochrosite is used as a source for manganese when it is available in large enough quantities. Sometimes banded masses of rhodochrosite are cut and used as ornamental stone or in large *objets d'art* where the markings can be displayed to their best advantage. Rhodochrosite crystals are usually found in hydrothermal veins and in sedimentary deposits. The term Inca-rose is derived from its formation as stalactites in abandoned mines which the Incas worked for silver.

The name rhodochrosite is derived from the Greek work *rhodon*, meaning pink, as crystals are normally a pink, faded pink or pinky-orange colour. Chemical alteration can, and will, turn crystals brown or black. Crystals, which display a translucence and a vitreous to pearly lustre, develop a dull oxidizing film of manganese when exposed to air. Rhodochrosite blackens gradually when heated.

**Range:** Rhodochrosite is quite a common mineral with important deposits occurring in Argentina, Mexico, Namibia, Romania, South Africa, Spain, the USSR and the United States.

*TOP LEFT An attractive example of deeply coloured rhodochrosite from Africa.*

*TOP RIGHT This vivid blue piece of azurite is from Zaire.*

# Azurite *(or Chessylite)*

Azurite was formerly used as a source of azure (blue) pigment. Due to its softness it is not very useful as an ornamental stone; it is, however, a copper ore of marginal importance. Azurite is a secondary copper mineral which occurs in sulphide deposits associated with carbonate rocks. It forms at lower temperatures than malachite, which often replaces it by ion exchange which occurs in an aqueous environment.

Azurite crystals tend to be azure-blue (hence the name) in colour, but can also be dark-blue with a green hint. Azurite crystals have a vitreous lustre, are soluble in ammonia and effervesce in dilute acids. Crystals will fuse easily, first turning black as they give up their water. Powdered azurite will, with time, turn greenish as it alters to malachite.

**Range:** The name chessylite is derived from Chessy, the name of the town in France where spherical aggregates are found. Azurite deposits also occur in Australia, Chile, Greece, Iran, Mexico and Namibia.

**Mohs' Hardness** *3½ – 4½*

**Specific Gravity** *3.3 – 3.7*

**Crystal Structure Hexagonal**

**Ornamental, Industrial**

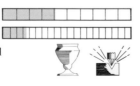

**Mohs' Hardness** *3½ – 4*

**Specific Gravity** *3.7 – 3.9*

**Crystal Structure Monoclinic**

**Industrial**

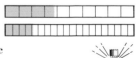

## Chalcopyrite *(or Copper Pyrite)*

Chalcopyrite ore accounts for almost 80 per cent of the world's copper metal. It is fairly common and widespread in its distribution, and often yields gold and silver as a by-product. Chalcopyrite occurs as a sulphide in high-temperature hydrothermal veins.

Chalcopyrite crystals are a brass colour, but over time tarnish to a range of colours (often with an iridescent film) depending on the environment and the chemical composition of the original crystal and the host rock. They are opaque with a metallic or sub-metallic lustre and will burn, colouring the flame green, and giving off a highly unpleasant gas.

**Range:** Substantial chalcopyrite deposits occur in Canada, Chile, Cyprus, the USSR (the Urals), the United States (Arizona, Montana and Utah), Zaire and Zambia. The Bingham Canyon copper mine in Utah is reputed to be the largest man-made hole in the world. The task of reclaiming the massive open-pit – due to commence once extraction had ceased – was going to be a major undertaking; that will now no longer be necessary as the area has been granted 'monument' status and attracts thousands of visitors a year.

*TOP LEFT Chalcopyrite, or copper pyrite, is the most common ore of copper. This sample is from Cumberland, UK.*

*TOP RIGHT An aggregate of dolomite crystals from Tsumeb, southern Africa.*

## Dolomite

Dolomite is an important building material, which is used as a structural and ornamental stone and for producing special cements. It is used in the manufacture of magnesia for refractories, as a metallurgical flux for the iron and steel industry, as well as being used in the chemical industry as a source of magnesium. Dolomite occurs most commonly as a sedimentary deposit believed to be caused by calcite alteration. It is typically deposited in low-temperature hydro-thermal veins, or is formed due to the partial metamorphism of dolomitic limestone into marble.

Dolomite is usually a collection of small, colourless, white, pale grey, pinkish or yellow-tinted crystals, frequently found in a saddle-shaped aggregate of slightly curved crystals. Dolomite fragments will dissolve with difficulty in hydrochloric acid and are not fusible. Dolomite resembles calcite in appearance and form, but calcite will effervesce and dissolve readily in cold hydrochloric acid.

**Range:** Thick beds of sedimentary dolomite occur throughout the world. The finest quality crystals come from Brazil (Bahia), Canada (Quebec), Italy, Switzerland and the lead-zinc mining area of the United States (Missouri).

**Mohs' Hardness** *3½ – 4*
**Specific Gravity** *4.1 – 4.3*
**Crystal Structure Tetragonal**
**Industrial**

**Mohs' Hardness** *3½ – 4*
**Specific Gravity** *2.85 – 2.95*
**Crystal Structure Hexagonal**
**Ornamental, Industrial**

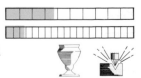

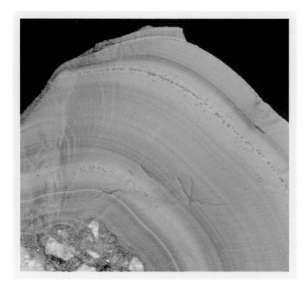

# Malachite

Malachite is used as an ornamental stone which – when cut into slabs and polished – is often made into boxes, small tables and *objets d'art* where its distinctive banding can be shown to its full advantage. Malachite, despite its exotic and distinctive markings, has rather humble origins; it is a secondary ore occurring in the upper levels of copper deposits, where it has been altered by the action of carbonated water.

It occurs in microcrystalline masses, usually as nodules with radiating bands. Malachite ranges in colour from weak green to emerald-green and from deep, dark green to blackish-green. It is fragile, has good cleavage and holds a silky polish well. It fuses fairly easily, colouring the flame green due to its copper content. It turns dark green and then black as it gives up its water, finally leaving a blob of metallic copper. Malachite crystals are rare and usually twinned.

**Range:** The most important sources of malachite in the form of banded masses are in Zaire, Zambia, Zimbabwe and the USSR (Siberia and the Urals). Less important sources occur in Australia, Chile, France, Namibia and the United States (Arizona and New Mexico).

MALACHITE

**M**alachite is thought to be a calming stone, its ripple-like markings seem to radiate out from nowhere. It is reputed to stimulate passive resistance and inner awareness. In ancient times it was thought to have the power to strengthen teeth, aid those with poor eyesight and warm a cold heart.

*TOP LEFT An example of malachite from Zaire which clearly shows both the concentric banding and the botryoidal form.*

*TOP RIGHT Sphalerite takes on a great variety of colours; this black example from Cumberland, UK, has been coloured by iron particles.*

# Sphalerite *(or Zinc Blend)*

The name Sphalerite is derived from the Greek word meaning 'treacherous', possibly due to the variations which commonly occur in its appearance. Sphalerite is a major source of zinc (used in alloying and galvanizing); it provides cadmium, gallium and indium as by-products. Sphalerite is usually associated with hydrothermal activity, and it occurs frequently with barite, challopyrite, fluorite and galena. Sphalerite sometimes occurs in sedimentary deposits or in low- and medium-temperature deposits where replacement crystal substitution has taken place.

Sphalerite varies in colour from yellowish-brown to reddish-brown (when pure), to blackish-brown when iron is present. It can also be green, pink, pinkish-orange (called 'honeyblend'), red (sometimes incorrectly called 'ruby-red') or colourless. Iron-rich sphalerites have a submetallic lustre while other variations have a resinous lustre. It is soluble in hydrochloric acid and infusible if pure, but will become more fusible as the iron content increases.

**Range:** Gem-quality sphalerite is found in Mexico and Spain, while less important deposits occur in England, Sweden, the United States (Kansas, Missouri, Oklahoma, Tennessee, Virginia) and Yugoslavia.

**Mohs' Hardness** *3½ – 4*
**Specific Gravity** *3.75 – 4.00*
**Crystal Structure Monoclinic**
**Ornamental**

**Mohs' Hardness** *3½ – 4*
**Specific Gravity** *3.9 – 4.2*
**Crystal Structure Isometric**
**Industrial**

# Celestite

Celestite is a principle source of strontium, which is used in the making of fireworks and signal flares as it produces a bright crimson colour when powdered and burned. It is used as an additive to battery lead and in the manufacture of rubber and paint. Other uses are found for it in the nuclear industry, in the refining of sugar beet, and in the preparation of iridescent glass and porcelain. Celestite is primarily a hydrothermal material which occurs as a late intrusion to cavities and voids in sedimentary limestone.

Celestite crystals are colourless, white, or white with irregular bluish zoning. The first crystals found were celestine blue (hence the name). They can also be yellow, reddish or brown tinted. Crystals tend to be squat and tabular, and often radiate from a base in a spiky aggregate. Crystals are transparent to translucent and display a pearly lustre. They are slightly soluble in water or acids and fuse easily.

**Range:** Crystals of exceptional quality have been found in England, Madagascar, Sicily, Tunisia and the United States (Put-in-Bay and Strontian Island on Lake Erie).

*TOP LEFT This spiky aggregate of celestite was found in Madagascar, a primary source of high-grade celestite crystals.*

*TOP RIGHT Crystals of cerussite are found in oxidized lead rich loads.*

# Cerussite

Cerussite is an important lead ore which sometimes contains recoverable silver. Quality crystals are sometimes cut into interesting gemstones which are modestly priced. It is found in the oxidation zones of lead-zinc deposits where it has been altered by the action of carbonated fluids acting on galena, and is often associated with unaltered galena and sphalerite.

Cerussite, which is often found as twin crystals, ranges in colour from colourless to black and can also be yellow, grey, or brown. crystals are uneven, often forming a grid-like aggregate network. They are brittle, turn brown when heated, dissolve with effervescence in nitric acid, and have a high specific gravity for a non-metallic mineral.

**Range:** Quality crystals occur in Australia, Austria, Czechoslovakia, Germany, Namibia, the USSR (Siberia), the United States (Colorado and New Mexico) and Zimbabwe.

**Mohs' Hardness** *3 – 3½*
**Specific Gravity** *3.96*
**Crystal Structure Orthorhombic**
**Industrial**

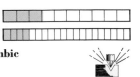

**Mohs' Hardness** *3 – 3½*
**Specific Gravity** *6.46 – 6.60*
**Crystal Structure Orthorhombic**
**Gemstone, Industrial**

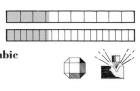

# Calcite

*TOP LEFT Sharp formations of colourless calcite from Cumbria, UK.*

*TOP RIGHT Columnar crystals of barite from Durham, UK.*

# Barite *(or Barytes)*

Compact masses of calcite, such as cement, lime, structural and ornamental stone, are used in the building industry. It is also used in metallurgical flux, fertilizers, fertilizer manufacture, and in the chemical industry. Until recently, clear calcite crystals were used as prisms. Occasionally rough crystals are used in jewellery or as ornaments. Calcite is formed by the evaporation of solutions rich in calcium salts. Such formations are stable under metamorphic conditions when they simply re-crystallize.

Calcite crystals are usually colourless or transparent, slightly milky, pearly white or pinky-white. They can be transparent, with a vitreous or pearly lustre. Calcite will dissolve in cold hydrochloric acid with a brisk effervescence.

**Range:** While calcite occurs worldwide, outstanding crystals have come from Czechoslovakia, Iceland and the United Kingdom.

Barite ore is the main source for Barium. Due to its relative inertness and lack of hydrogen ion concentration (pH), its main use is as a weighting additive in drilling and production fluids, which are used in the oil and gas retrieval industries. It is also used in conjunction with radiography to diagnose digestive blockages; when mixed with cement it is used as a screen against radiation. Further uses for it are found in the manufacture of paper or white pigments. Barite usually occurs as a low- to medium-temperature hydrothermal vein, as a sedimentary deposit in veins and cavities, or as an evaporate. When it occurs in limestone it is often associated with lead or silver.

Barite varies in colour from colourless to brown or bluish, and can be white, yellowish or red tinted. It is soft, non-metallic and can occur as tabular or columnar crystals, or in a rosette form known as 'desert rose'. It is insoluble in acids and fuses with difficulty, colouring the flame yellow-green.

**Range:** 'Desert rose' crystals are common in the Sahara (North Africa) and in Oklahoma (United States). Impressive barite crystals are also found in Czechoslovakia, England and Romania.

**Mohs' Hardness** *3*
**Specific Gravity** *2.71*
**Crystal Structure Hexagonal**
**Industrial, Ornamental, Gemstone**

**Mohs' Hardness** *2½ – 3½*
**Specific Gravity** *4.48 – 4.5*
**Crystal Structure Orthorhombic**
**Industrial**

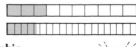

# Chalcocite
### (or Redruthite or Copper Glance)

The name redruthite comes from the name of the town Redruth in Cornwall (England) which was the main source of the world's copper and tin industry for much of the last century. Chalcocite is an important source of copper metal and is associated with hydro-thermal copper-sulphide deposits as a secondary-enriched zone.

Chalcocite rarely occurs as crystals, but when it does, they are soft, malleable, pseudo-hexagonal and striated. Chalcocite more typically occurs as a granular, dull-grey aggregate, whose surface has altered to green or black. It is fusible and dissolves easily in nitric acid.

**Range:** Quality chalcocite crystals have been found in Chile, England, Mexico, Namibia, South Africa, Spain and the United States (Connecticut and Montana).

*TOP LEFT A granular mass of chalcocite with some bornite, from Cornwall, UK.*

*TOP RIGHT A crystalline formation of copper (also called native copper) from Canada.*

# Copper (or Native Copper)

After iron, copper is probably the most important metal in man's history. Used as an alloy for centuries, it is now the most important metal in the field of electrical engineering. Copper is typically formed by reduction in oxidization zones of sulphide deposits, or, in mined-out sites where copper-sulphated water will precipitate onto iron objects, or replace organic fibres, such as wood.

Copper is copper-red when freshly broken, but tarnishes rapidly to a dull, dusty looking brown colour. Crystals are soft, have a metallic lustre, and fracture in an irregular manner. Copper rarely occurs as crystals; it is found more frequently in compact masses or branch-like shapes. Copper is very malleable, and dissolves easily in nitric acid, staining the resulting solution an azure-blue colour.

**Range:** The finest crystals of copper come from the United States (Michigan). Less spectacular examples come from deposits in Chile, Germany, Mexico, Spain, Sweden, the USSR and Zambia.

**Mohs' Hardness** *2½ – 3*
**Specific Gravity** *5.5 – 5.8*
**Crystal Structure Orthorhombic**
**Industrial**

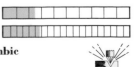

**Mohs' Hardness** *2½ – 3*
**Specific Gravity** *8.93 – 9.00*
**Crystal Structure Isometric**
**Industrial**

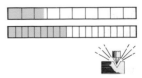

# Galena

Galena is the richest, most common, and most important ore for lead. Even so, many lead mines are only economical due to their silver deposits, which, more often than not, are found in them as well. Lead is used in batteries, in soldering (due to its low melting point), as an additive to petrol to retard ignition, and as a constituent of paints. Galena, which usually occurs in cubic form (or flattened and drawn-out when subjected to post-growth pressures and movement) is associated with calcite veins contained in limestone or dolomitic masses.

Galena crystals are dull grey and cubic in shape. They have a warm feeling, will leave a grey mark on the hands if rubbed, and are heavy. They are soluble in heated hydrochloric acid, and give off an unforgettable 'bad egg' smell.

**Range:** Important lead deposits occur in Australia, England, the United States (Idaho and Missouri) and West Germany.

*TOP LEFT Cubic crystals of galena on limestone; this example is from Derbyshire, UK.*

*TOP RIGHT These rare crystals of silver were found in Mexico, one of the main sources of fine silver formations.*

# Silver

Known to the Incas as the 'tears from the moon', silver is a precious metal which can occur in crystal form. Silver and its derivatives are used in industry as a chemical catalyst; in electronics, because of its high conductivity; and in dentistry, due to its inertness. It is also used in jewellery. By far the largest use for silver is in the field of photography, because of the sensitivity of silver bromide to light.

Silver crystals are rare, but when they do occur, they are associated with hydrothermal veins; there, they are wire-like, soft, malleable, and silver-white when freshly broken. They tarnish rapidly to brown, dull-grey or black. Silver, however, more generally occurs in irregular masses as silver associated with gold and/or copper.

**Range:** The finest wire-like crystals come from Norway; they are also found in East Germany and Mexico. Important silver deposits are found in Australia, Canada, Chile and the United States (Colorado, Nevada and South Dakota).

**Mohs' Hardness** *2½ – 3*
**Specific Gravity** *7.2 – 7.6*
**Crystal Structure Isometric**
**Industrial**

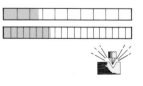

**Mohs' Hardness** *2½ – 3*
**Specific Gravity** *10.5 pure*
**Crystal Structure Isometric**
**Industrial, Ornamental**

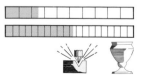

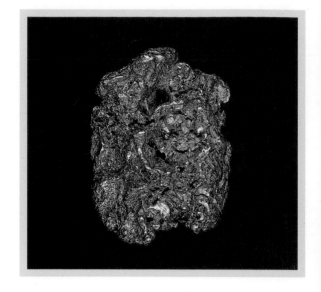

# Brucite

Brucite is widely used as a refractory material in the extraction of magnesia and as a source of metallic magnesium and its salts. It is a metamorphic mineral found in low-temperature serpentized rocks or in contact metamorphosed dolomites. They are also found as an alteration product in the final stages of a metamorphic incident.

Brucite commonly occurs as scaly, plate-like or highly fibrous aggregates. It is transparent or translucent, being either colourless, white, yellowish or occasionally pink. It is infusible but dissolves readily in cold diluted acid with no effervescence. It parts easily along micaceous cleavage planes, which then display a dullish-pearly lustre. Brucite is often an associated mineral of asbestos.

**Range:** Brucite occurs in Canada, Scotland, the USSR and the United States (Nevada, Pennsylvania, and Texas).

*TOP LEFT A botryoidal mass of brucite found in Africa.*

*TOP RIGHT This 1 inch (2 cm) gold nugget was mined in the Urals, USSR.*

# Gold

Known to the Incas as the 'Sweat of the Sun', gold is a rare precious metal which can occur in crystal form. It has been fashioned into ornaments and jewellery since the dawn of time. During the last millenium gold has become established as a way of storing wealth, and as a hedge against the economic effects of natural disaster, war and man's folly. In more recent times, gold's ability to conduct heat and electricity without ever tarnishing has led to it being used widely in technology and industry.

Though primarily of hydrothermal origin, large concentrations of gold are formed by the erosion and re-deposition of gold-bearing lavas. Gold crystals are golden-yellow when pure ('yellow gold') tending to silvery-yellow when alloyed with silver ('white gold') and reddish-orange when mixed with copper ('red gold'). The most distinguishing features of gold are its weight and the fact that it will never tarnish or rust.

**Range:** Of the major gold-producing areas in the world, the Witwatersrand district in the Transvaal (South Africa) and the Carlin Trend in Nevada (United States) are the most famous. Australia, Brazil, Canada and Papua New Guinea also contain major gold deposits.

**Mohs' Hardness** *2½*

**Specific Gravity** *2.3 – 2.5*

**Crystal Structure Hexagonal**

**Industrial**

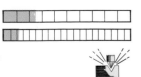

**Mohs' Hardness** *2½*

**Specific Gravity** *19.3 pure*

**Crystal Structure Isometric**

**Ornamental, Industrial**

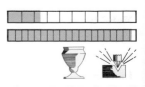

# Halite
## *(or Common Salt or Rock Salt)*

*TOP LEFT A group of halite crystals showing perfect cubic development.*

*TOP RIGHT A fine example of the pseudo-hexagonal form often taken by muscovite. This piece is from the USA.*

# Muscovite
## *(or Common Mica or White Mica)*

Halite has been used for thousands of years as a means of preserving meat. It is still used today for this purpose and in food preparation in general. Its most important market is in the chemical industry for manufacturing soda, sodium and hydrochloric acid.

Halite is predominantly an evaporate, often aided by precipitation in enclosed seas. Such deposits are overlaid often by bands of clay and shale; however, being plastic in nature and having a low density, halite tends to flow upwards (like a bubble enclosed in thick honey). The resulting 'salt domes' are often worked for their halite or sulphur, or for the hydrocarbon traps which often occur at the edges of the salt dome and the host rock.

Halite crystals are commonly colourless, white, yellow or red. They have a vitreous lustre, are brittle, have a pleasant salty taste and over a period of time will dissolve in the water that they have absorbed from the atmosphere.

**Range:** Large halite deposits occur in Austria, Czechoslovakia, England, France, Germany, Poland, the USSR (Siberia) and the United States (Louisiana and Texas).

Muscovite is a superb electrical and heat insulator and is used for this purpose either in small sheets, or is powdered and re-formed with cement and plastic. It is used in the oil industry as a plugging material when fluids are being lost to porous or cavernous formations; as a filler in paper and rubber manufacture; and as an addition to badly drained soils.

Muscovite commonly occurs in silica- and aluminium- rich igneous or metamorphic environments. It is extremely common in sand and other sedimental materials – which are formed as a result of the disintegration and weathering of igneous and quasi-igneous rocks – where it has crystallized from gases.

Muscovite is characterized by its perfect cleavage and the resulting flexible flakes. Crystals are commonly pseudo-hexagonal, forming 'mica blocks' within a matrix of sheets displaying no apparent structure. It can vary in colour from colourless to black and can be also white, yellow, reddish or brown.

**Range:** Huge single crystals (measuring eight yards across) have been found in Brazil, Canada (Ontario), India and the United States (New Hampshire and South Dakota).

**Mohs' Hardness** *2½*
**Specific Gravity** *2.1 – 2.2*
**Crystal Structure Isometric**
**Industrial**

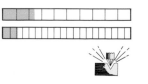

**Mohs' Hardness** *2 – 3*
**Specific Gravity** *2.77 – 2.88*
**Crystal Structure Monoclinic**
**Industrial**

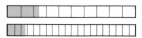

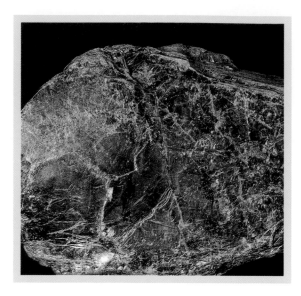

# Vermiculite

Vermiculite is a clay-form, which when heated rapidly, expands dramatically. Expanded, vermiculite is an excellent thermal and acoustic insulator and is used for this purpose in the building industry. In recent years demand for it has increased due to the gap left in the market when the health problems associated with asbestos were recognized. Vermiculite is also used as a filler in the paper, rubber and plastics industry, and as a packaging medium. It has recently gained much favour as an additive to heavy clay and badly irrigated soils because it can increase water percolation. Vermiculite is caused by the hydrothermal alteration of biotite and phlogopite.

Vermiculite is a plate-like, golden, or honey-coloured crystal which frequently occurs as a scaly aggregate. It has a pearly vitreous lustre, is slightly soluble in acids and expands by up to as much as 20-fold when heated beyond 572°F (300°C).

**Range:** The largest vermiculite deposit occurs in South Africa with other significant masses located in Argentina, Australia, Canada and the United States (Massachusetts, Montana and North Carolina).

*TOP LEFT A sample of vermiculite taken in the USA, showing clearly the pearly vitreous lustre of this crystal.*

*TOP RIGHT Argentite veins within a matrix of quartz; this example was found in Mexico.*

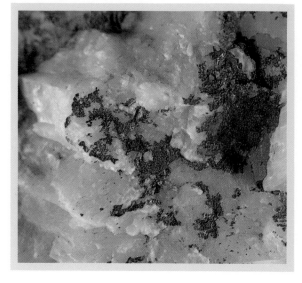

# Argentite *(or Silver Glance)*

Argentite is an important silver ore. It frequently occurs with galena from which it is separated by melting the ore and skimming the silver off the top where it tends to float. Silver has been used for centuries as a metal in jewellery and for plating purposes.

Argentite is frequently found in low-temperature hydrothermal veins with other silver minerals, usually as a massive agglomeration, either in groups of branching crystals or in distorted form. It is soft and heavy, and tarnishes quickly to a ghostly, grey-green, dull colour. Argentite is metallic and shiny when freshly cut or broken. It fuses easily into a silvery blob.

**Range:** Argentite crystals have been found in Bolivia, Czechoslovakia, Mexico and Norway. As a mineral deposit, Argentite is found in Australia, Canada, Chile, Peru and the United States (Colorado and Nevada).

**Mohs' Hardness** *2 – 3*
**Specific Gravity** *2.4 – 2.7*
**Crystal Structure Monoclinic**
**Industrial**

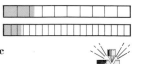

**Mohs' Hardness** *2 – 2½*
**Specific Gravity** *7.2 – 7.4*
**Crystal Structure Isometric**
**Industrial**

# Bismuth *(Native Bismuth)*

Bismuth is the main source of bismuth metal which is used in alloys with a low melting point, in lubrication additives and in medicines and cosmetics. It is a fairly uncommon mineral which is found in hydrothermal veins where it is often associated with cobalt, nickel, silver, and tin ores.

Bismuth crystals are rare but, when they do occur are usually imperfectly formed. Bismuth is very heavy, opaque with a bronze-coloured metallic lustre, and fuses easily at a low temperature to produce a metallic blob. Bismuth dissolves easily in nitric acid, which will then yield a white precipitate if diluted with water. Bismuth usually occurs in massive form or in a foliated aggregate.

**Range:** Substantial bismuth deposits occur in Bolivia. It can also be found as an associate mineral with lead, cobalt and silver in Canada, East Germany and Norway.

*TOP LEFT Although bismuth crystals occur naturally, this superb specimen is a by-product of smelting.*

*TOP RIGHT Borax, shown here in association with ulexite; this sample is from Iran.*

# Borax

Borax is used as a cleaning agent, in the manufacture of high-temperature glass (such as pyrex), and as a flux in soldering, brazing and welding. It is also used as a source of boron, which has the ability to absorb neutrons and is used in the control rods employed in nuclear reactors. Borax is an evaporate which has formed due to the evaporation of ponded-in saline lakes and seas.

Borax normally occurs as soft, prismatic, stubby crystals, which range in colour from colourless to white and from whitish-blue to yellowish-white. Colourless crystals tend to lose their water and assume a dusty white appearance not unlike square marshmallows. Borax is water soluble and has a slightly sweet, alkaline taste.

**Range:** In ancient times, borax was brought to Europe from the salt lakes of Tibet. Today, the main deposit is in the United States (Death Valley, California) with smaller deposits being found in Argentina and Turkey.

**Mohs' Hardness** *2 – 2½*
**Specific Gravity** *9.7 – 9.83*
**Crystal Structure Hexagonal**
**Industrial**
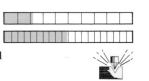

**Mohs' Hardness** *2 – 2½*
**Specific Gravity** *1.70 – 1.74*
**Crystal Structure Monoclinic**
**Industrial**

# Cinnabar

# Gypsum

Cinnabar is the most important source of mercury metal, and was used in the past as the mineral pigment known as vermillion (or vermilion). Cinnabar crystals are rare, but mostly occur as a coarse, granular, or compact aggregate which is found in masses and impregnations, in lavas, and near hot spring deposits. It also occasionally occurs in sedimentary deposits which have acted as a replacement medium from nearby igneous rocks.

Cinnabar is opaque, with a dull, pink, washed-out, earthy-red colour. It is soft, heavy, and deposits droplets of mercury on cold surfaces after being heated. It is insoluble in acids, but can be attacked by aqua regis and chlorine gas. Cinnabar fractures unevenly, is brittle and tends to splinter.

**Range:** For industrial purposes, the most important cinnabar deposits occur in Italy, Spain and Yugoslavia. Smaller deposits occur in Algeria, China, Peru, the USSR and the United States (Arkansas, California, Texas and Utah).

Gypsum is the most common sulphate mineral. Massive stratified gypsum deposits are worked in the Paris basin. Much of this is used in the manufacture of plaster of Paris, from which the name is derived. This remains the principle use for gypsum. In Europe powdered limestone is used as a filter to reduce sulphur gas emissions from coal-burning power-stations. This limestone is chemically changed into a type of 'dirty gypsum' which is unsuitable for fine plaster-work where whiteness is required, but it is still used in other forms of building work.

Gypsum is also used as a retarder in cement, as a fertilizer and as a flux in glass fabrication. The pyramids just outside Cairo were originally faced with alabaster (a form of gypsum).

Gypsum, which can be colourless, white, yellow, grey or brown, commonly occurs as elongated flattish crystals with a silky lustre. Crystals are frequently bent, and large crystals are not uncommon.

**Range:** Splendid gypsum crystals occur in Chile, Mexico, Sicily and the United States (Utah), while extensive deposits are mined in England, France (Paris), the USSR and the United States.

**Mohs' Hardness** *2 – 2½*
**Specific Gravity** *8.0 – 8.2*
**Crystal Structure Hexagonal**
**Industrial**

**Mohs' Hardness** *2*
**Specific Gravity** *2.35*
**Crystal Structure Monoclinic**
**Industrial**

# Stibnite

*TOP LEFT Stibnite crystals are often formed as striated columns radiating out of a base, as in this example from Romania.*

*TOP RIGHT Sulphur crystals, such as these found in Sicily, Italy, are associated with volcanic activity where they often occupy voids and crevices.*

# Sulphur *(or Native Sulphur)*

Stibnite is the main ore for antimony which is alloyed with tin and copper to produce an anti-friction alloy. It is also alloyed with the lead used in storage battery plates. Antimony is used in the manufacture of matches, as a colour generator in fireworks, and as an industrial pigment. The slats derived from stibnite are used in the vulcanizing of rubber, in medicine and in glass fabrication. Stibnite is a low-temperature mineral that often occurs in hydrothermal veins.

Stibnite crystals, which are soft and steely-grey, commonly occur as either spines radiating out from a base or as striated columns. They are soft, often bent, and have a brilliant metallic lustre. Thin splinters will fuse in a match flame, whilst powdered stibnite is soluble in concentrated hydrochloric acid. A few drops of this solution will precipitate bright orange in potassium iodine.

**Range:** Superb crystals, sometimes over a foot (30 cm) in length, have been found in China, Japan and Romania.

Sulphur is mostly used for the manufacture of sulphuric acid, which is then used to make fertilizers. It is also used for vulcanizing rubber, and in the manufacture of explosives and insecticide. Sometimes referred to as the 'Devil's milk' due to its distinctive smell, most of the sulphur produced today is as a by-product of hydrocarbon production and processing.

Large sulphur lenses do, however, occur at the top of salt domes in the Gulf of Mexico, off the coast of the United States. These deposits, which are often discovered when drilling for oil, are 'mined' using the Frasch process. Superheated water is pumped down a borehole into the formation, melting – but not dissolving – the sulphur. The 'melt' is then aerated with hot air bubbles (to keep it fluid) and lifted to the surface via a pipe set in the same borehole.

Pure sulphur is yellow, but is more likely to be beige, dull brown or black (especially in Iran). The darker the colour the higher the entrapped hydrocarbon impurities. Sulphur crystals are translucent and burn with a blue flame.

**Range:** The most prolific area of sulphur production is the United States (Louisiana and Texas) with sublimate deposits occuring in Chile, Japan and Sicily.

**Mohs' Hardness** *2*
**Specific Gravity** *4.6 – 4.7*
**Crystal Structure Orthorhombic**
**Industrial**

**Mohs' Hardness** *1½ – 2½*
**Specific Gravity** *2.00 – 2.10*
**Crystal Structure Orthorhombic**
**Industrial**

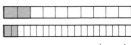

# Bauxite

Bauxite is one of the world's most commonly occurring ores. Its extraction is often dependent on the availability of cheap transport and/or cheap electricity. It requires cheap transport since bauxite mining is a victim of the 'economy of scales' syndrome; cheap electricity is required since this is needed to refine bauxite into aluminium. Bauxite is a secondary deposit: it is what remains after aluminium-bearing rocks have been weathered in tropical and subtropical climates.

Bauxite is commonly red-brown to dull brown, but can also be white, yellowish or grey in colour. It is dull, finely grained, often earthy in texture and appearance. Generally, it is not a very interesting or prepossessing mineral. Nonetheless, it is the world's main source of aluminium.

**Range:** Important bauxite deposits occur in Ghana, Hungary, Indonesia, Jamaica, Surinam and the USSR.

*TOP LEFT Bauxite, the world's major source of aluminium, is an agglomeration of several crystal forms. This piece was mined in France.*

*TOP RIGHT Graphite is best known as a component of lead pencils; this sample was found in Africa.*

# Graphite

Graphite is used in lead pencils, and is mixed with mineral oils or grease to form a high-temperature lubricant. It is also used in the form of electrodes, as brushes in electrical motors, and in protective paints. Graphite usually occurs in metamorphic rocks as the final stage in the carbonization of organic substances. Graphite is occasionally found as thin hexagonal crystals with triangular basal markings, but it occurs more frequently as foliated masses or thin sheets.

Graphite is steely-grey to black, or opaque with a sub-metallic dull lustre. It is greasy to the touch and very soft. Graphite is frequently used as one of the comparison materials in Mohs' hardness test kits. It will form a grey-black streak on paper. Graphite displays perfect basal cleavage which results in thin platelets. It is insoluble in acid and will only melt in extremely high temperature.

**Range:** Significant deposits of graphite are found in Czechoslovakia, Madagascar, Mexico, South Korea, Sri Lanka and the USSR.

**Mohs' Hardness** *1 – 3*
**Specific Gravity** *2.00 – 2.7*
**Crystal Structure Aggregate of various crystal types**
**Industrial**

**Mohs' Hardness** *1 – 2*
**Specific Gravity** *2.09 – 2.23*
**Crystal Structure Hexagonal**
**Industrial**

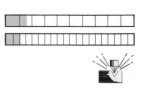

# Molybdenite

Molybdenite is the major ore of molybdenum and is used extensively for special-purpose alloys and as a dry lubricant whose performance is unaffected by high temperatures. It is formed in very high-temperature, igneous environments, and is one of the earliest metallic minerals to crystallize out of volcanic melts. It occurs typically in pegmatite dykes, veins within certain granites, or in dolmitic rocks. It occurs some-times in contact with metamorphic deposits, usually near limestones.

Molybdenite crystals are rare, but when they do occur are lead-grey or bluish-grey in colour and not dissimilar to graphite or galena in appearance. Molybdenite more commonly occurs as foliated or scaly aggregates. These are easily separated along perfect cleavage planes into flexible platelets. Molybdenite is greasy to the touch, opaque, and with a lustre which is sometimes metallic in character. It can also appear to be tired and dull.

**Range:** Molybdenite occurs in Canada, Chile, China, Norway, the USSR and the United States (Colorado).

*TOP LEFT A very metallic-looking example of molybdenite from Scotland, UK.*

*TOP RIGHT Montmorillonite, shown here swelling rapidly on being placed in water. This example is from Wyoming, USA.*

# Montmorillonite *(or Bentonite)*

Montmorillonite is the chief constituent of the clay mineral bentonite. Montmorillonite absorbs water – especially water that has a high alkaline value – and swells accordingly. A liquified clay like this is used in the oil drilling and construction industries due to its ability to suspend solids. It can also seal the pores of formations which normally accept fluids. Montmorillo-nite will become semi-solid when left undisturbed, but will become pumpable again when agitated (this is called being thixotropic).

Montmorillonite is also used as a purifying medium and as a filler in the manufacture of paper and rubber. It is formed either in a hydrothermal environment where volcanic ash has been altered, or in a sedimentary tropical environment where feld-spars have been altered. It is a microcrystalline material which is white, grey, or beige in colour. It forms earthy masses which are greasy to the touch and crumble easily.

**Range:** Montmorillonite is found in large masses in Montmorillon in France (hence the name), Germany, Japan and the United States (Alabama, California and Florida).

**Mohs' Hardness** *1 – 1½*
**Specific Gravity** *4.6 – 5.0*
**Crystal Structure Hexagonal**
**Industrial**

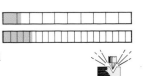

**Mohs' Hardness** *1*
**Specific Gravity** *1.2 – 2.7*
**Crystal Structure Monoclinic**
**Industrial**

# Talc *(Soapstone)*

Talc is usually powdered and used as a filler in paper, rubber and paints. It is also used in the textile and cosmetics industry. Slabs of talc are used to acid-proof laboratory surfaces, while larger pieces are often fashioned into simple statues. Talc is a secondary mineral formed at the metamorphic alteration of any one of a number of magnesium silicates.

Talc never occurs as visible, individually formed crystals, but as a white, grey or greenish, pearly, folia-ted compact mass. It is greasy to the touch (hence the name soapstone) and is used as the lowest of the comparison materials in Mohs' hardness scale. Talc is insoluble in acid and almost impossible to fuse.

**Range:** The largest talc mine in the world is in the Pyrenees mountains (France), where high-quality talc is still selectively mined by hand in what is a primitive but highly profitable operation. Due to the altitude and weather, the open-cast mine is operated only for six months a year. Smaller deposits occur in Austria, Australia, Canada, India, Korea and South Africa.

*TOP LEFT Talc or soapstone, such as this piece from the USA, is commonly used in powdered form in many industries.*

*TOP RIGHT Spheres of mercury, the only metal which is liquid at room temperature, shown here on a sample of cinnabar.*

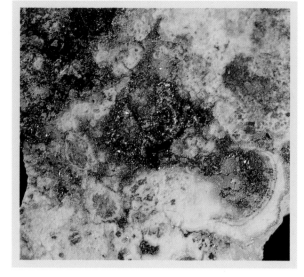

# Mercury

Mercury is used in mineral processing for recovering gold and silver. It is also used in the manufacture of explosives, batteries and electrical rectifiers. During the last century, mercury was used for curing beaver furs which were made into hats. The vapours given off were eventually found to be poisonous and to cause insanity. Mercury usually occurs in association with cinnabar in areas of volcanic or geyser activity.

Mercury is the only metal which is liquid at room temperature. It usually occurs in nature as tiny red beads which appear to be weeping out of bright red cinnabar. Its liquid nature, its association with cinna-bar, and its weight are distinctive features of the metal. Mercury will dissolve in nitric acid.

**Range:** The most important mercury deposits are found in Italy, Spain and Yugoslavia.

**Mohs' Hardness** *1*

**Specific Gravity** *2.58 – 2.83*

**Crystal Structure Monoclinic**

**Industrial, Ornamental**

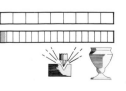

**Mohs' Hardness** *zero – at room temperature*

**Specific Gravity** *13.6*

**Crystal Structure Hexagonal**

*— below minus 39°C*

**Industrial**

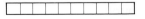

# *Glossary*

**Acicular.** Crystals which display a very long, needle-like habit. Often radiating from a base.

**Aggregate.** An assembled mass of more than one crystal type and usually more than one mineral species.

**Alluvial.** Material which has been transported by water and accumulates to form a rock mass.

**Aqua Regis.** A mixture of one part of hydrochloric acid to one part of nitric acid. Capable of dissolving gold.

**Asterism.** An optical effect which appears as a star-shape within a crystal. Caused by minute secondary crystal inclusions.

**Authigenic.** A mineral formed within a sedimentary environment and not one which has been introduced.

**Birefringent.** An optical phenomenon which causes any image viewed through such a crystal to have a double appearance.

**Carat.** A unit of weight used for gemstones. One carat is equal to 3 grains (200 mg).

**Cleavage.** The plane along which a crystal will naturally break. This may not necessarily be a well defined plane.

**Columnar.** A crystal shape which resembles a column; a regular elongated prism.

**Concretion.** A concentration of mineral growth around a nuclei within a sedimentary environment.

**Contact.** The transmission of temperature from an igneous rock, without a pressure change; often leads to a metamorphic change.

**Dendrital.** The skeleton form of a crystal usually found on fracture surfaces; capable of absorbing matter.

**Density.** Ratio of the weight of an object to its volume. Frequently referred to as the specific gravity.

**Deposit.** An accumulation of minerals in sufficient quantity and quality so as to be economically extractable.

**Dike.** An igneous intrusion of great length but limited thickness which often fills a vein or fracture plane.

**Ductile.** Bends easily, usually without any permanent damage (like warm liquorice sticks).

**Eruptive.** Being ejected from below the earth's surface; lava, dust, ashes, gases and vapours.

**Evaporite.** A mineral of chemical origin which owes its formation to the evaporation of an aqueous solution.

**Extrusive.** When molten rocks or lava flow out of the earth's surface.

**Fluorescence.** The temporary emission of light waves to give colours not normally seen.

**Foliated.** The crystal habit that resembles foliage (hence the name). Minerals are usually flaky in character.

**Fracture.** To break unevenly. This occurs in crystals which do not have a clearly defined cleavage.

**Gangue.** That part of a mineral which is of no value but which must still be extracted.

**Gel.** A semi-solid solution, usually heavily charged with elements; often solidifies to form coloured minerals.

**Geode.** Spherical cavity; much favoured by crystals as a location to grow in.

**Greisen.** Igneous rocks which have been altered by fluids, rich in volatile elements.

**Habit.** The characteristic shape of a crystal; the form in which it most frequently occurs.

**Hardness.** The resistance to wear. Often measured by means of the Mohs' scratch test scale.

**Hydrothermal.** The process by which heavily charged aqueous solutions transport and form minerals.

**Igneous.** A rock which is formed by the solidification and crystallization of molten rocks or magma.

**Inclusion.** The entrapment within a growing crystal of a gas, a liquid or a solid (often as a secondary crystal growth).

**Leaching Zone.** The area in the lower part of a mineral deposit, where chemical compounds are leached and then redeposited.

**Mafic.** The characteristic of a mineral which refers to its predominance of ferromagnesium minerals.

**Magmatic.** The formation of minerals from molten silica-rich rocks, contained within the earth's surface.

**Malleable.** Soft material (usually refers to metals) which can be formed into shape without the need to be beaten.

**Metamorphism.** The transformation of a rock from one state to another; caused by the effects of a nearby heat and pressure source.

**Mineral.** A naturally formed homogeneous solid which possesses a clear chemical compostion.

**Oxidation.** The chemical process by which oxygen is added to a compound or process. Opposite to reduction.

**Pegmatites.** Igneous intrusions caused when residual liquids cool from magmas; gives an ideal growing region for crystals.

**Piezoelectric.** Ability of a crystal to emit an electrical charge if compressed, or vice-versa.

**Pipe.** A volcanic structure, usually tube-like, through which magmatic material is forced upwards.

**Pleochroism.** The ability of a mineral to reflect and absorb different colours, depending on their direction of orientation.

**Reduction.** The chemical process by which oxygen is removed from a compound or process. The opposite to oxidation.

**Schists.** Metamorphic rocks which contain mineral depositions (usually mica-like) in parallel or subparallel veins.

**Synthetic.** A precious mineral reproduced in a laboratory, with the same characteristics as that of a natural stone.

## Picture Credits

l = left; r = right; c = centre; t = top; b = below.
British Museum; page 17 br.
British Museum (Natural History): pages 12, 14, 23 r, 41 l, 42 r, 47 l, 63 r.
C.M. Dixon, Canterbury: page 17t.
Geoscience Features Picture Library/A. Fisher: pages 22 r, 25 l, 25 r, 26 r, 27 r, 28 l, 34 l, 35 r, 37 l, 37 r, 38 l, 39 l, 40 l, 40 r, 46 l, 49 r, 52 l, 56 l, 56 r, 57 l, 58 l, 59 l, 59 r, 60 l, 60 r, 61 l, 62 r, 63 l, 64 l, 65 l, 65 r, 67 r, 68 l, 73 l, 76 r.
Geoscience Features Picture Library/Dr B. Booth: pages 7, 22, l, 23 l, 24 l, 24 r, 26 l, 27 l, 28 r, 29 l, 29 r, 30 l, 30 r, 31 l, 31 r, 33 l, 33 r, 34 r, 35 l, 36 l, 36 r, 38 r, 39 r, 41 r, 42 l, 43 l, 43 r, 44 r, 45 l, 45 r, 46 r, 47 r, 48 l, 48 r, 49 l, 50 l, 51 r, 52 r, 53 l, 53 r, 54 l, 54 r, 55 l, 55 r, 57 r, 61 r, 62 l, 66 l, 66 r, 67 l, 68 r, 69 l, 69 r, 70 l, 70 r, 71 l, 71 r, 72 l, 72 r, 73 r, 74 l, 74 r, 75 l, 75 r, 76 l.
Geoscience Features Picture Library/W. Hughes: pages 17 bl, 18.
Nature magazine/Dr Stephen H. Richardson: page 13 t. Reprinted by permission from Nature vol. 310 19 July 1984; copyright © 1984 Macmillan Magazines Ltd.
Peter Nixon (University of Leeds): pages 6 t, 8, 9, 13 c, 13b, 19 b, 19 t, 32 l, 32 r, 44 l, 50 r, 51 l, 58 r, 64 r.
Oriental Museum, University of Durham: page 6 b.
Cover pictures courtesy of Geoscience Features Picture Library.

# Index

Italic page numbers refer to illustrations

## A

acicular crystals 77
adamantine lustre 15
adularia moonstone 12, 47, 47
agate 7, 39, 39
  quartz cat's-eye 36, 36
aggregate structure 77
  bauxite 74, 74
alabaster 72
alexandrite 14, 23, 23
alluvial material 77
almandine 37, 37
aluminium 74
amazonite (amazon stone) 45, 45
amethyst 33, 33
andalusite 28, 28
antimony 73
apatite 14, 56, 56
aqua regis 77
aquamarine 19, 25, 25, 27
argentite 70, 70
asbestos 15
asterism 77
authigenic minerals 77
axe stone 41
axinite 40, 40
azurite 61, 61

## B

barite 65, 65
barium 65
barytes 65, 65
bauxite 15, 74, 74
benitoite 46, 46
bentonite 75, 75
Bernhardi, Reinhard 10
beryl group
  aquamarine 25, 25
  beryl 9, 15
  bixbite 26, 26
  emerald 26, 26
  golden beryl 27, 27
  morganite 27, 27
  pink beryl 27, 27

  precious beryl 25, 25
  red beryl 26, 26
birefringence 14, 30, 34, 39, 77
bismuth (native bismuth) 71, 71
bixbite 26, 26
bloodstone 49, 49
blue zoisite 42, 42
bonamite 59, 59
borax 71, 71
boron 71
brazilianite 52, 52
bronzite 53
brucite 68, 68

## C

cadmium 63
cairngorm 35
calamine 57
calcite group
  calcite 14, 65, 65
  inca-rose 61, 61
  rhodochrosite 61, 61
californite 43
Cape ruby 30, 30
carat 77
cassiterite 15, 43, 43
categorization 10–12
cat's-eye 23
  quartz cat's-eye 36, 36
celestite 64, 64
cerussite 64, 64
chalcedony 40, 40
chalcocite 66, 66
chalcopyrite 62, 62
chemical impurities 13, 13
chessylite 61, 61
chiastolite 28
chromian diopside 54
chrome 52
chromite 52, 52
chrysoberyl 23, 23
cinnabar 72, 72, 76
citrine 33, 33
cleavage 15, 77
clinozoisite 44
cobalt 60
colour 13–14

colourless quartz 34, 34
columnar crystals 77
concretion 77
constancy of interfacial angles 10
contact 77
copper 18, 61, 62, 66
copper (native copper) 66, 66
copper glance 66, 66
copper pyrite 62, 62
cordierite 30, 30
corundum group
  corundum 14
  ruby 13, 22, 22
  sapphire 13, 23, 23
cristobalite 50
crocidolite 36
crystal engineering 18
crystal formation 7, 9, 19
  crystal solutions 9
  growth rate 9
crystal habit 12
Curie, Marie and Pierre 55
cyprine 43

## D

danburite 29, 29
dendrital crystals 77
density 77
deposit, meaning 77
desert rose 65
diamond 9, 13, 14, 15, 19, 19, 22, 22
dichroite 30, 30
dike 77
diopside 54, 54
dioptase 56, 56
disthene 58, 58
dolomite 62, 62
dry-bone ore 59, 59
ductile materials 77

## E

emerald 14, 19, 26, 26
enstatite 53, 53
epidote 44, 44
eruptive materials 77
evaporite minerals 77

extrusive materials 77

## F

fairy stones 31, 31
feldspar group
  adularia moonstone 12, 47, 47
  amazonite (amazon stone) 45, 45
  labradorite 46, 46
  moonstone 47, 47
  prthoclase 48, 48
  spectrolite 46, 46
flame-fusion process 18–19
fluorescence 77
fluorite (fluorspar) 14, 60, 60
foliated minerals 77
fool's gold 48
fracture 77

## G

galena 67, 67
gallium 63
gangue 77
garnet group
  almandine 37, 37
  Cape ruby 30, 30
  garnet 13, 17
  grossular 23, 37, 37
  pyrope 30, 30
  rhodolite 38, 38
  spessartite 38, 38
gel 77
geode 7, 77
gold 17, 62, 68, 68, 76
golden beryl 27, 27
graphite 74, 74
green tourmaline 32, 32
greisen rocks 77
grossular 23, 37, 37
gypsum 14, 15, 72, 72

## H

habit, meaning 78
hafnium 39
halide group 60, 60
halite 7, 69, 69
hardness 14–15, 78
hematite 17, 49, 49

hemimorphite 57, 57
hexagonal structure 11
hiddenite 44, 44
hornblend 36
hornstone 41, 41
hydrostatic balance 16, 16
hydrothermal action, crystals formed by 9, 78

## I

idocrase 43, 43
igneous rocks 78
inca-rose 61, 61
inclusion 78
indicolite 29, 29
indium 63
industrial uses 17–18
  acid-resistant products 58
  alloys 53, 57, 63, 66, 71, 73, 75
  batteries, manufacture of 64, 67, 73, 76
  bearings 22
  catalysts 58, 67
  ceramic products 28, 48, 58, 64
  chemical industry 56, 65, 69
  conductors 58, 67, 68
  construction industry 62, 65, 70, 72, 75
  cosmetics 71, 76
  dentistry 48, 67
  drilling, boring, grinding and cutting 22
  electrical engineering 66, 69
  electrical rectifiers 76
  electrodes 74
  electronics industry 58, 67
  explosives, manufacture of 73, 76
  fertilizers, manufacture of 48, 56, 65, 72, 73
  fireworks and signal flares 64, 73
  food preparation 69
  glass manufacture 64, 71, 72, 73
  hydrofluoric acid, production of 60

insecticides, manufacture
of 73
insulators 28, 48, 58, 69, 70
lubricants 71, 74, 75
matches, manufacture of
73
medicine 7, 55, 65, 71, 73
metal ores 17
metallurgical flux 62, 65,
71
nuclear industry 55, 64, 71
packaging 70
paint and pigments 17, 49,
51, 52, 61, 64, 65, 67, 72,
73, 74, 76
paper manufacture 65, 69,
70, 75, 76
pencils, manufacture of 74
petrochemical industry
60, 65, 67, 69, 75
photography 67
plastics industry 60, 70
polishing 22, 49
refactories 28, 52, 62, 68
rubber industry 64, 69, 70,
73, 75, 76
soldering 67, 71
sugar beet, refining 64
sulphuric acid,
manufacture of 48, 73
textile industry 76
welding 71
iolite 30, 30
iron 49
isometric structure 11

J

jade 41, 41
jasper 41, 41

K

kunzite 45, 45
kyanite 15, 58, 58

L

labradorite 46, 46
lapis lazuli (lapis) 17, 17, 51,
51
German (Swiss) 41
lavrovite 54
lazulite 51, 51, 54, 54
leaching zone 78
lead 64, 67
light, natural and artificial

limonite 17
lithium 45
lodestone 49, 49
lustre 15

M

mafic minerals 78
magmatic minerals 78
magnesia 62, 68
magnesium 62, 68
magnetite 49, 49
malachite 61, 63, 63
malleable materials 78
manganese 61
mercury 72, 76, 76
metamorphism 78
mica 69, 69
milky quartz 9, 34, 34
mineral, meaning 78
Mohs' scale 14–15, 78
molybdenite 75, 75
monoclinic structure 11
montmorillonite 15, 75, 75
moonstone 47, 47
morganite 27, 27
morion 35

N

native bismuth 71, 71
native copper 66, 66
native sulphur 73, 73
navigational aids 30, 49
nephrite 41
nickel 53, 60
nickeline (niccolite) 53, 53
noble orthoclase 48
non-crystalline structure
50, 50

O

olivine 42, 42
opal (precious opal) 7, 50,
50
orthoclase 14, 48, 48
noble 48
orthorhombic structure 11
oxidation 78

P

pegmatites 78
peridot 42, 42
phenacite (phenakite) 28,
28

phosphorus 56
piemontite 44
piezoelectric crystals 78
pinl beryl 27, 27
pipe structures 78
pistacite 44, 44
pitchblende 55, 55
plaster of Paris 17, 72
platinum 58, 58, 60
pleochroism 45, 78
prehnite 47, 47
pycnometer 16, 16
pyrite 48, 48
pyrolusite 17
pyrope 30, 30
pyroxene group
diopside 54, 54
enstatite 53, 53
pyrrhotite 60, 60

Q

quartz group
agate 39, 39
amethyst 33, 33
chalcedony 40, 40
citrine 33, 33
colourless quartz 34, 34
hornstone 41, 41
jasper 41, 41
milky quartz 9, 34, 34
opal 50, 50
quartz 10, 14, 15
quartz cat's-eye 36, 36
rock crystal 34, 34
rose quartz 35, 35
smoky quartz 35, 35
smoky topaz 35, 35
tiger-eye 36, 36

R

radium 55
red beryl 26, 26
redruthite 66, 66
reduction process 78
relative density 16
rhodochrosite 50, 61, 61
rhodolite 38, 38
rhodonite 50, 50
rock crystal 34, 34
rock salt 7, 17, 69, 69
rose quartz 35, 35
Rosiwal scale 15
rubellite 31, 31

ruby 13, 14, 18–19, 22, 22

S

salt 7, 17, 69, 69
sapphire 7, 13, 14, 19, 23, 23
water sapphire 30
scheelite 57, 57
schists 13, 78
schorlite 32
scientific uses
enstatites 53
prehnite 47
pyrrhotite 60
uraninite 55
sekaninaite 30
siberite 31
silver 62, 64, 67, 67, 70, 76
silver glance 70, 70
smithsonite 57, 59, 59
smoky quartz 35, 35
smoky topaz 35, 35
soapstone 6, 76, 76
sodalite 51, 51
specific gravity 16
spectrolite 46, 46
specularite 49, 49
spessartite 38, 38
sphalerite 63, 63
spinel 24, 24
spodumene group
hiddenite 44, 44
kunzite 45, 45
star signs
Gemini 25
Leo 22
Libra 22
Scorpio 24
Taurus 26
staurolite 31, 31
Steno, Nicolaus 10
stibnite 73, 73
Strasser, Joseph 18
striations 12
strontium 64
sulphur (native suphur) 15,
73, 73
sun stones 33
synthetic crystals 18–19, 78

T

talc 14, 76, 76
tanzanite 42, 42

tetragonal structure 11
thixotropic substances 75
thorium 39
tiger-eye 36, 36
tin 43
topaz 14
precious 24, 24
smoky 35, 35
tourmaline group
green tourmaline 32, 32
indicolite 29, 29
rubellite 31, 31
tourmaline 19, 32, 32
verdelite 32, 32
triclinic structure 11
trigonal structure 11
tungsten 57
turquoise 17, 55, 55

U

uraninite 55, 55
uranium 55
utahlite 59, 59

V

variscite 59, 59
verdelite 32, 32
vermiculite 70, 70
Verneuil, A.V.L. 18
vesusianite 43, 43
Vikings
lodestone 49
Vikings' compass 30
violane 54
vitreous lustre 15

W

water sapphire 30
white mica 69, 69
wilulite 43

X

xanthite 43

Z

zinc 57, 59, 63
zinc blend 63, 63
zircon 19, 39, 39
zirconium 39
zoisite, blue 42, 42